T0252388

An Accidental
Statistician

An Accidental Statistician

THE LIFE AND MEMORIES OF GEORGE E.P.BOX

George E.P. Box

Professor Emeritus
Department of Statistics
University of Madison—Wisconsin
Madison, Wisconsin

With a little help from my friend, Judith L. Allen

Copyright © 2013 by John Wiley & Sons, Inc. All rights reserved

Published by John Wiley & Sons, Inc., Hoboken, New Jersey
Published simultaneously in Canada

No part of this publication may be reproduced, stored in a retrieval system, or transmitted in any form or by any means, electronic, mechanical, photocopying, recording, scanning, or otherwise, except as permitted under Section 107 or 108 of the 1976 United States Copyright Act, without either the prior written permission of the Publisher, or authorization through payment of the appropriate per-copy fee to the Copyright Clearance Center, Inc., 222 Rosewood Drive, Danvers, MA 01923, (978) 750-8400, fax (978) 750-4470, or on the web at www.copyright.com. Requests to the Publisher for permission should be addressed to the Permissions Department, John Wiley & Sons, Inc., 111 River Street, Hoboken, NJ 07030, (201) 748-6011, fax (201) 748-6008, or online at http://www.wiley.com/go/permission.

Limit of Liability/Disclaimer of Warranty: While the publisher and author have used their best efforts in preparing this book, they make no representations or warranties with respect to the accuracy or completeness of the contents of this book and specifically disclaim any implied warranties of merchantability or fitness for a particular purpose. No warranty may be created or extended by sales representatives or written sales materials. The advice and strategies contained herein may not be suitable for your situation. You should consult with a professional where appropriate. Neither the publisher nor author shall be liable for any loss of profit or any other commercial damages, including but not limited to special, incidental, consequential, or other damages.

For general information on our other products and services or for technical support, please contact our Customer Care Department within the United States at (800) 762-2974, outside the United States at (317) 572-3993 or fax (317) 572-4002.

Wiley also publishes its books in a variety of electronic formats. Some content that appears in print may not be available in electronic formats. For more information about Wiley products, visit our web site at www.wiley.com.

Library of Congress Cataloging-in-Publication Data:

Box, George E. P.
  An accidental statistician : the life and memories of George E.P. Box / George E. P. Box.
    pages cm
  Includes index.
  ISBN 978-1-118-40088-3 (cloth)
1. Box, George E. P. 2. Statisticians–United States–Biography. I. Title.
  QA276.157.B69A3 2013
  519.5092–dc23
  [B]
                                        2012040251

10 9 8 7 6 5 4 3 2 1

This book is dedicated to my students, with whom it was my privilege to work, and who became my friends.

Contents

Foreword

$\mathcal{V}$IRGINIA WOOLF wrote about a character with a mind that "kept throwing up from its depths, scenes, and names, and sayings, and memories and ideas, like a fountain spurting over." George Box is the embodiment of that active mind. Dinner with George is a spurting of stories, poems, songs, and anecdotes about his work and his friends. *An Accidental Statistician* jumps you into that fountain of ideas. It is great fun even if books about statistics and science are normally absent from your reading list.

No doubt many readers think, as I once did, that the subject is difficult and dull. Here we have a charming and colorful storyteller who quotes Yogi Berra in a discussion on the analysis of variance; has Murphy, of Murphy's Law fame, ringing the alarm when there is an opportunity to make things better; and explains an experiment with critical variables named "banging" and "gooeyness." There are stories about composite designs, time-series forecasting, Evolutionary Operation, intervention analysis, and so on, but they are not mathematical, and most include personal anecdotes about people who were involved in their invention and original application. You learn about statistics and science and, simultaneously, meet a literal Who's Who of statisticians and scientists, and the Queen of England as well.

I met George Box in 1968 at the long-running hit show that he called "The Monday Night Beer Session," an informal discussion group that met in the basement of his house. I was taking Bill Hunter's course in nonlinear model building. Bill suggested that I should go and talk about some research we were doing. The idea of discussing a modeling problem with the renowned Professor Box was unsettling. Bill said it would be good because George liked engineers. Bill and several of the Monday Nighters were chemical engineers, and George's early partnership with Olaf Hougen, then Chair of Chemical Engineering at Wisconsin, was a creative force in the early days of the newly formed Statistics Department. I tightened my belt and dropped in one night, sitting in the back and wondering whether I dared take a beer (Fauerbach brand, an appropriate choice for doing statistics because no two cases were alike). I attended a great many sessions over almost 30 years, during which hundreds of Monday Nighters got to watch George execute an exquisite interplay of questions, quick tutorials, practical suggestions, and encouragement for anyone who had a problem and wanted to use statistics. No problem was too small, and no problem was too difficult. The output from George was always helpful and friendly advice, never discouragement. Week after week we observed the cycle of discovery and iterative experimentation. We saw real examples that, "All models are wrong, but some are useful." We saw how statistics is a catalyst for scientific method, and how scientific problems catalyze ideas for doing statistics. What a treat.

My business is water quality engineering. One night I wanted to discuss a problem that involved a measurement called the Biochemical Oxygen Demand (BOD). George asked whether it would be all right for him to explain this BOD test. He gave as good an explanation as I ever heard. I asked how he happened to know that and learned that at age 16, he took a job as an assistant chemist in a sewage treatment plant. One year before I was born, in 1939, at age 19, he published a paper about oxygen demand in the activated sludge wastewater treatment process, which at the time was new and poorly understood. George's paper can stand with papers on the subject written by some famous Wisconsin engineers who worked on the problem at about the same time. In the 1990s, 55 years later, George and I worked on forecasting the dynamics of activated sludge process performance using multivariate nonstationary time series. Imagine that from a world-famous statistician who was one of the earliest researchers on this widely used wastewater treatment process.

George and I have one bit of unfinished piece of business. A 20-foot-tall civil war soldier guards the stone arch entrance to Fort Randall Park, which is next to the engineering building and the statistics building at the University of Wisconsin—Madison. George had thought for some time that the soldier should have a medal. In 1993, we found a suitable brass medal in a sidewalk stall near Hyde Park, London, but our plan to hang it around the soldier's neck was never fulfilled. I now believe that the medal should stay with George as an award of merit for memoir writing. He deserves it. (And we two old friends do not have to climb the soldier.)

Last night, May 10, 2012, my wife and I had dinner with George and Claire at their house in Shorewood Hills. He said, "The memoir is finished." I asked, "What's your next project. It's hard to picture you not doing some writing every day." He answered, "I'm thinking of a paper about Fisher's idea on multiplicative effects in experiments."

An Accidental Statistician is finished, but apparently George is not. That is good news. Thank you, George.

P. Mac Berthouex
Emeritus Professor
University of Wisconsin

Second Foreword

$\mathcal{I}$T is a pleasure to welcome this autobiography of an extraordinary scholar and gentleman: George Edward Pelham Box. My intersection with George Box's long and active statistics career begins in the summer of 1952, when I was a research associate at the Aberdeen Proving Grounds, MD, working for Dr. Frank Grubbs. Dr. Grubbs and I together read George Box's wonderful paper on response surface experimentation,[1] and I recall how we both marveled at its simplicity ... obvious of course now that the message was offered! Originality and elucidation are the signs of Box's genius. On returning that September to my graduate studies at the Institute of Statistics, NC State, in Raleigh, I learned that George Box had agreed to come to Raleigh for a year as a visiting Research Professor. In January 1953, I became his first graduate student.

It's a pleasure to read about George's boyhood. His father worked hard to provide a modest family environment in a society that offered advantaged youths the greatest opportunities. We learn how George, through the good fortune of meeting alert teachers, uncovered his talents as a writer and more modestly in mathematics. He starts his young adult life as an assistant chemist collecting data on wastewater treatment, and it is here that his career as a statistician begins. And what a career it proved to be!

One of George Box's distinguishing characteristics is that he only occasionally published a paper as a sole author. This memoir introduces us to a host of his students and research associates, and it provides colorful vignettes of these many wonderfully varied personalities who became his co-authors. Some ten years ago, a group of his students gathered together a compendium of his papers covering the statistical fields of

---

[1]"On the Experimental Attainment of Optimum Conditions" Box, G. E. P. and Wilson, K. B. (1951) *Jour. Royal Stat. Soc. Series B*, <u>13</u>, 1–45.

quality, experimental design, control, and robustness.[2] Both neophytes and savants have found the exposition within these many papers superb. Of course, there are also his co-authored textbooks on experimental design,[3] time series analysis,[4] Bayesian inference,[5] and control,[6] wherein elucidation of the subject's theory and application repeatedly prove both original and illuminating. For George Box, the acronym "KISS" translates into "Keep It Sophisticatedly Simple."

I found one recollection of George Box's early statistical experiences particularly fascinating. Near the end of World War II, the British came into possession of German explosive shells containing unknown deadly gases. George was part of the small group that first determined the spectacular deadliness of tiny concentrations of these new reagents (actually nerve gases). They were never employed in warfare, but had portions been dropped over major cities, the population consequences could easily have rivaled those of Hiroshima and Nagasaki. To have been present working along that thin boundary keeping the world from additional disaster is impressive.

We learn, too, of those whom George Box considered *his* mentors and hear tales of his early collaborations with others well known among the statistical fraternity. And throughout this autobiography, we become aware of Box's broad contributions to modern statistical theory and practice. His papers and books have vastly expanded awareness of Bayesian methods and time-series modeling. We find the production of information-laden data to be a statistical specialty that enhances scientific progress as it moves from initial conjecture through experimentation and data analysis, leading on to new conjecture. And beyond statistically formal matters, we also capture his deep appreciation for the mind's ability, or better still that of a collection of minds, to give birth to new models and not usual

---

[2] *Box on Quality and Discovery*: George Tiao, Søren Bisgaard, William J. Hill, Daniel Peña, Stephen M. Stigler. (2000) John. Wiley & Sons.

[3] *Statistics for Experimenters*: Box, G.E.P, Hunter, William G. and Hunter, J. Stuart (1978) John Wiley & Sons.

[4] *Time Series Analysis: Forecasting and Control*: Box, G.E.P and Jenkins, Gwilym M. (1970) Holden Day.

[5] *Bayesian Inference in Statistical Analysis*: Box, G.E.P. and Tiao, George C. (1973) Addison Wesley.

[6] *Statistical Control by Feedback and Adjustment*: Box, G.E.P. and Lucño, Alberto (1997) John Wiley & Sons

conjectures. It seems that George Box's advice to all those pondering a problem is to be sure to think out of the box.

J. Stuart Hunter
Professor Emeritus
Princeton University

Preface

HERE is a story about a very tall man who was walking with his four-year-old son to pick up a newspaper. He suddenly realized that his son was having difficulty keeping up with him. He said, "Sorry, Tommy, am I walking too fast?" And the boy said, "No, Daddy, I am."

Now this account can be viewed in two ways: as an amusing story, or joke or, as illustrating the essence of scientific discovery. The boy's view of the situation was correct but not obvious. The father's view was obvious but wrong.

So it is perhaps no coincidence when humor and scientific insight come together. Good science is a form of wit, of seeing the joke that nature is playing on us.

At 93, I can look back on quite a few examples.

The Box family circa 1895. From left clockwise, Uncle Bertie, my grandfather and grandmother, my father, Aunt Daisy, Uncle Pelham, and Aunt Lina.

Acknowledgments

$\mathcal{F}$OR much of the time that this memoir was being written, I have been quite ill. This has placed an enormous burden on my wife, Claire, who herself has been ill during much of this time. Had it not been for her devoted help, so generously given, this memoir could never have been written. This help has been realized by a practical, ingenious, and well-trained mind. Whenever there is a crisis, she has not only known what to do, she has done it with cheerful understanding and expertise.

In addition, I am especially grateful to:

Bovas Abraham	Judy Hunter
David Bacon	Stu Hunter
Ford Ballentyne	Brian Joiner
Ernesto Barrios	Tim Kramer
Mac Berthouex	Kevin Little
Sue Berthouex	Alberto Luceño
Claire Box	Merve Muller
Joan Box	Vijay Nair
Helen Box	Lars-Erik Öller
Harry Box	Judy Pagel
Robin Chapman	Daniel Peña
Norman Draper	José Ramírez
Conrad Fung	Marian Ros
Larry Haugh	Xavier Tort
Margaret Homewood	John Sølve Tyssedal

Brent Nicastro for his permission to use various photographs.

And to Judith Allen, my friend.

Acknowledgments

From The Publisher

To those of you who do not know George Edward Pelham Box well, suffice it to say that he is a titan in the field of statistics. He is a self-taught statistician who utilized his experience and knowledge of statistics to create unique contributions to many areas particularly in process improvement. And, he is a nice guy, to boot. He rarely—if ever—needs an introduction. His very presence is our present.

This book is being published in the year of George's ninety-fourth birthday as a memoir of his life, his friends, and his contributions. We know that, during his academic tenure, he wrote over 2000 journal articles; published twelve books for Wiley alone (see list below) resulting in over a quarter-of-a-million copies sold worldwide; and was responsible for helping to get *Technometrics*, a joint publication of the American Society for Quality and the American Statistical Association, off the ground.

We also know from first-hand experience that George is a true gentleman, a loving father, and a dedicated husband. He has influenced the lives, in no small way, of everyone he has touched, from young aspiring statisticians to experienced editors-in-chief. When you see a grin on his face, you know that he is about to espouse a bit of wisdom mixed-in with a tad of advice and always with a joke. He most often accomplishes what he sets out to achieve, without fanfare or accolade. He is probably the most unforgettable character a graduate student or editor-in-chief has ever had the experience and pleasure to know.

Two people have assisted George in the production of this book. They include his loving wife of twenty-seven years, Claire Box, and his friend and research assistant, Judith Allen, both of whom supported him as he wrote this book.

The management and staff of Wiley commend Dr. Box for all that he has done to enrich the world of statistics, both here and abroad. We wish him continued "presence" and the peace of mind that he will always remain a titan in the written word and in our hearts for generations to come.

Box Titles Published with Wiley

Evolutionary Operation: A Statistical Method for Process Improvement	Box-Draper, 1969
Statistics for Experimenters: An Introduction to Design, Data Analysis, and Model Building	Box-Hunter-Hunter, 1978
Empirical Model-Building and Response Surfaces	Box-Draper, 1986
Bayesian Inference in Statistical Analysis	Box-Tiao, 1992
Statistical Control: By Monitoring and Feedback Adjustment	Box-Lucēno, 1997
Evolutionary Operation: A Statistical Method for Process Improvement	Box-Draper, 1998
Statistics for Experimenters: Design, Innovation, and Discovery, 2nd Edition	Box-Hunter-Hunter, 2005
Improving Almost Anything: Ideas and Essays, Revised Edition	Box-Friends, 2006
Response Surfaces, Mixtures, and Ridge Analyses, 2nd Edition	Box-Draper, 2007
Time Series Analysis: Forecasting and Control, 4th Edition	Box-Jenkins-Reinsel, 2008
Statistical Control by Monitoring and Adjustment, 2nd Edition	Box-Lucēno-Paniagua Quiñones, 2009
An Accidental Statistician: The Life and Memories of George E.P. Box	Box, 2013

"Who in the world am I? Ah, that's the great puzzle."[1]

CHAPTER ONE

Early Years

GRAVESEND, where I was born, is about 25 miles east of London on the River Thames. The river there is about a mile across, and at that time ships from all over the world came by on their way to the London docks. As a ship would come up the river, three tugs would hurry across to its side and accompany it while it moved on. From the first of these, you could watch the pilot climb aboard; then, from the second, the health officer; and finally the customs officer. There were occasionally large vessels coming from the Far East, Australia, New Zealand, or India that could not travel further, and so they were moored in the middle of the river. Thus, Gravesend was very much concerned with the sea, and people such as pilots, lightermen, and customs and health officials abounded (Figure 1.1).

My grandfather, also named George, was a grocer and an "oil and color" merchant—that is, one who sold paint. My father, Harry, was the youngest boy. My father's oldest brother, whom we called "Uncle Bertie," attended private school and took degrees in theology and semitic languages at Oxford. He became a rector, wrote a number of esoteric and scholarly books, and was rarely heard from again.

In 1892, Pelham, the second son, was lured under false pretenses to make his fortune in the United States. He was about 20 when he got off a train in Nebraska to find nothing but the howling wind, but

---

[1] All quotations appearing above chapter titles are from Lewis Caroll, *Alice's Adventures in Wonderland*, originally published in 1865 by Macmillan.

An Accidental Statistician: The Life and Memories of George E.P. Box, First Edition. George E.P. Box.
© 2013 John Wiley & Sons, Inc. Published 2013 by John Wiley & Sons, Inc.

FIGURE 1.1
Pilot leaving Terrace Pier, Gravesand, Original by Anthony Blackman.

he later returned and stayed in the United States, becoming a citizen and eventually working for the railroad in Chicago. When he retired, he moved to Florida where he had a small citrus grove.

Over the years, my family fell on increasingly difficult times. My father had hoped to go to engineering school, but by the time he was a young man, the family had little money and he had few career choices. Two sons had already left, so my father stayed and found work as a "clothier's assistant." He had a hard life. When I was growing up, he still worked in a tailor shop, at Tilbury docks, across the river from Gravesend. To get to work from our house on Cobham Street, he had to walk about a mile to the Town Pier at the bottom of High Street, cross on the ferry, walk some more, get on a train that took him to Tilbury docks, and then walk again to get to the shop. In the evening, he would face the same journey in reverse, sometimes in the pouring rain. He was poorly paid—two pounds ten shillings a week was barely a living wage.

Because people had to use coal for heating and cooking, the resulting fog could sometimes be so thick that objects four or five feet away were invisible. My father got across the river then in a small boat. On foggy days, lightermen (who transferred goods between ships and docks) made extra money by taking people across.

When I was about nine, I learned to use stilts, and walking on these, I would meet my father at the end of our road. He would sometimes have a pennyworth of roasted chestnuts in a paper bag that he would share with me.

From the time I was about five years old, I sometimes went with him to Tilbury. I liked perching myself at the very front of the ferry and watching it cut through the water. My father had a friend called Mr. Launder who kept a tobacconist and barber shop at Tilbury docks. Mr. and Mrs. Launder, who did not have any children, liked for me to come and visit them. I provided some entertainment to the people waiting at the barber shop by reciting poetry. I remember one poem that began, "Great Wide Beautiful Wonderful World," and a line I liked was "World, you are beautifully dressed." I enjoyed poetry and tried writing some myself.

Despite his hard life, my father was a happy man. With the help of my sister Joyce, he would frequently organize picnics and parties. Our parties were not like parties now. There were no alcoholic drinks. (It wasn't that we were teetotalers; we just didn't have any money.) We gathered around the piano and sang sentimental Victorian songs; most of these sound pretty silly now. We also played all kinds of party games: musical chairs, hunt the slipper, "murders," and so forth. In addition, we performed plays of our own invention. And there were mysteries when my father demonstrated the power of the magic wand.

When we wanted to go somewhere for a picnic, we walked. Cars were for rich people. Although we did not have a car, we did have a "barrow," which we pushed. This was my father's invention. It could be steered with two pieces of rope, and in the back, it carried supplies: cricket bats and picnic things. We would walk with the barrow to some pretty place we liked, perhaps about three or four miles away.

The countryside was lovely. Cobham village was only four miles away, and the church and the two pubs there were old, unspoiled, and beautiful. One pub was called the Dickens Inn because Charles Dickens had written some of his books there, and he chose this locale for some of his stories. In his book *Great Expectations* (1860, Chapman & Hall), you will find, for example, that the prison ship from which the convict

Magwitch escaped was just below Gravesend, and it was near Gravesend, that Magwitch was finally caught. Close at hand inland was the village of Meopham (pronounced "Meppam"), with a fine cricket field, and a lovely place for picnics called the Happy Valley. There we sat in the grass, made a fire for cooking and heating water for tea, and we sometimes played cricket.

My mother had a difficult life, with such meager resources, trying to provide for a family. She fed and cared for not only our immediate family but also for various relatives who lived with us while I was growing up. And she also had to contend with my father, whose kindness and generosity knew no bounds. I am afraid her life was often one of quiet desperation.

But it was not all gloomy. Sometimes we went to the park on Windmill Hill, where we would play and my parents would have a beer at the pub that backed up to the park. On other occasions, my mother and my father would go out together in the evening, perhaps to see a movie, and my sister Joyce would take care of my brother and me (Figure 1.2).

Joyce was ten years older than me. Her mother, my father's first wife, had died in an influenza epidemic. Joyce could always come up with an exciting game, and we often played cowboys and Indians. Joyce had a lot of responsibility within the family but was a very good sport about it. In addition to watching over my brother and me, she was also my father's chief helper around the house. My father was a wonderful handyman, and this was fortunate because we had so little money. He could do most things that others hired people to do: roof mending, gas fitting, paper hanging, and so on. But every now and again, he and Joyce quarreled and Joyce would run downstairs very upset and say, "I won't help him anymore!" After about ten minutes, my father would come downstairs and apologize. Then they would make up and things would go along quietly for a bit (Figure 1.3).

In 1926, a man called Alan Cobham became famous when he flew from England to Australia and back, landing his plane in the River Thames in front of the Houses of Parliament. He was promptly knighted. Later, Sir Alan formed Cobham's Flying Circus, a group of fliers that went around England showing off their skills and giving plane rides to the public.

FIGURE 1.2
Growing up, Clockwise from left to right, me, my brother
Jack, and my sister Joyce.

In the early 1930s, Cobham's Flying Circus was performing in a field close to Gravesend and Joyce and I went to watch. Plane rides were part of the show, and to my astonishment, Joyce said, "Come on, Pel, let's go up in an airplane!" I said, "But Joyce, it's *five shillings*!" Five shillings seemed a huge amount of money. "Just this once," she answered, and so up we went in this small plane and thrilled to the views of town and countryside. Although it wasn't a very long flight, it was my first and I never forgot it.

FIGURE 1.3
My father and Joyce.

Joyce worked at Woolworth's, where she eventually became an overseer. I realized much later how she had contributed substantially to the family budget for many years and, in particular, how her sacrifice had made it possible for my brother Jack and me to get an education. I know if she had had the chance, she would have done just as well.

My brother Jack was three years older than me. He could not have been more than 13 when he became very interested in amateur (ham) radio. He designed and built a transmitter, and as he couldn't apply for a ham radio license until he was 17, for some years, he had a "pirate" station. The town had many ham radio enthusiasts, and they used to communicate with others throughout the world. When you made a contact, you exchanged "QSL cards." These were decorated postcards showing your amateur station's designated call sign in large letters (G6BQ in my brother's case). You might be particularly proud, for example, of a QSL card from some remote part of China. The cards were often used by hams to decorate the walls of their "shacks." Jack used a small shed that my father had constructed at the back of the house for his shack. Thinking about it now, my father, sister, and I formed a cohesive group. Jack was always happy with his radio (Figure 1.4).

Behind the houses on Cobham Street, there was an extensive area of allotments. These were rectangular plots of land that you could rent for growing vegetables. They also proved useful for erecting antenna poles to achieve long-range radio contact. Erecting an antenna pole was quite a task, so friends and other hams came and helped. Reception depended to some extent on the direction in which the antenna was pointing, so Jack managed to persuade a number of allotment holders to allow poles to be erected on their plots so that he had antennas facing in all directions. From the nearby railway station, you could see the various antennas, and stories got around, especially during the Cold War, about

FIGURE 1.4
Jack's radio.

their purpose. I remember one ham radio colleague complaining to me about the number of Jack's poles. He said, "I don't mind helping him with *one* pole, but he wants them all over the place."

There was a scheme to take over the allotments to build a car park for the many people who took the train to London. This plan didn't suit Jack, and for weeks he went around collecting signatures opposing the car park, arguing disingenuously that the allotments were needed to grow food. His scheme worked for a number of years, but I have seen in a recent photograph that the car park is now firmly established.

When I was about ten I came a cross a book called *The Boy Electrician*.[2] The fascinating thing about it was that it was a "can do" book. The apparatus and experiments that the author described could all be constructed from components which were readily available. He told for example how to make an electric bell or burglar alarm, a morse telegraph, an experimental wireless telephone and an electric motor.

From the first day I saw it, the book was seldom in the library. I had a friend, Jim Tatchol, who was equally interested and we spent hours together making, or trying to make, the things in the book. In doing so we learned a great deal. We were also fortunate in having a very sympathetic physics teacher who after school spent hours with us helping to make things work.

When I was about eleven years of age, I used *The Boy Electrician* to build myself a crystal set. With this I would listen to the BBC on headphones after I went to bed. From about 10 p.m. until midnight, the station transmitted live dance music from one or another of the big hotels in London. They all had dance bands, one of the most famous being the Savoy Hotel Orpheons. Others were named for the dance bandleaders: Jack Payne, Harry Roy, Geraldo, and so on. I can still sing many of the songs that were popular then. My antenna was looped around the ceiling, and I could get good reception provided that my brother was not on the air.

The establishment of the BBC as an independent entity was largely due to its first director, Sir John Reith, to whom we remain forever grateful. Sir John, however, was very religious, so on Sundays, the station carried only religious programs. On those days, we listened instead to Radio Luxemburg and Radio Paris, which were commercial stations that

---

[2]Alfred P. Morgan, *The Boy Electrician*, Lathrop, Lee & Shepard, 1913

carried special programs in English. I can still remember the jingles that accompanied the commercials, such as:

We are the Ovaltineys,
Little boys and girls;
Make your request, we'll not refuse you,
We are here just to amuse you.
Would you like a song or story?
Will you share our joys?
At games and sports we're more than keen;
No merrier children could be seen,
Because we all drink Ovaltine,
We're happy girls and boys!

[CITATION: Harry Hemsley, "We Are the Ovaltineys," 1935; theme song on Radio Luxembourg show, "The Ovaltineys Concert Party" from 1935 to 1940.]

My family lived in a rather large, semidetached house at 52 Cobham Street, which as I mentioned, we often shared with relatives and friends. When I was small, there was Mr. Strickland, a lodger, who lived downstairs in the basement. My mother told me that as a child I walked in my sleep and on one occasion had walked down three-and-a-half flights of stairs and said to Mr. Strickland, "Go away, Stricky, I'm in the middle of a dream." After Mr. Strickland died, we took over the two basement rooms for a kitchen and dining room.

As a child, my best buddy was my maternal grandmother. I used to sit on her bed, and she told me stories and read to me. This was when I first heard what is still my favorite book, *Alice in Wonderland*. Except for the kitchen, the rest of the house was not heated, but there was always a fire in Grandma's room and I used it to make hot buttered toast for both of us. When she died and I couldn't find her, an aunt told me that she had "gone to live with Jesus." I said, "I don't want her to live with Jesus. I want her to live with me."

Aunt Lina, my father's sister, lived in a room on the ground floor of our house when I was about ten years old. (She is pictured at lower right in the 1893 Box family photo locates in the Preface.) Aunt Lina read a lot, but she was stone deaf. I could only communicate with her by writing things down, and she enjoyed this. I must have realized how lonely her life was because I taught both of us sign language, and from then on, we communicated with our hands. One game that greatly amused her was when the family was gathered around the table for a meal. I would sign a funny message with my fingers, such as "Uncle So and So makes a terrible noise when he's drinking soup," and she would go into fits of laughter. The joke was, of course, that normally she was entirely cut off from what was being said, but *this* message only she and I could understand.

After I went to grammar school, I had a friend, Cyril Jones, whose parents had a car. One day we drove down to a place called Tudely-cum-Caple near Dover to see Aunt Daisy, who was my father's other sister. She and her husband had a small holding that did not look very prosperous. Soon after this, her husband died and Aunt Daisy came to live with us. She also was deaf, but not totally, and she had a very primitive hearing aid. She liked to dance, and I was frequently recruited as a partner (Figure 1.5). Like my father, she was a happy person and liked to play the piano and sing.

And then there was Uncle Willy, who was actually my father's first cousin. He lived 12 miles away in Gillingham and, to my mother's dismay, would turn up unexpectedly and expect to be fed. He had money, which we didn't, so we did our best not to offend him. He had worked for the Admiralty on airship design and had helped design the airship R33, which crashed. He had also been on a trip up the Amazon River, and he had composed a long lecture, "One Thousand Miles Up the Amazon," which he illustrated with lantern slides. He endlessly rehearsed this at our house, consulting my father on the text. I can remember my father coming home from a hard day's work and my mother's cry of despair, "Uncle Willy's here." He was very mean with his money, and I recall him telling my mother how he had once saved a penny by changing buses on his way to see us.

Uncle Willy eventually died, and we inherited his musical instruments, which included a banjo, a guitar, a violin, an organ, and a player piano. My father could produce a tune from almost any musical instrument, but what he liked best was what we called the organ, which was really a souped-up harmonium.

Willy had been a keen photographer, so we also inherited a very nice set of lenses. As a boy, I very much wanted a camera, but we didn't have any money to buy one. I saw an advertisement for a new newspaper,

FIGURE 1.5
Dancing with Aunt Daisy.

The Daily Herald, that was to begin publication. To encourage circulation, they printed a coupon with each issue and after you had collected 100 consecutive coupons, you could get a free camera. My father was already receiving a different paper, the *Daily Chronicle*—we called it the *Daily Crocodile*—which was more aligned with his political views. With his typical kindness, he switched to the new paper (although he much preferred the other), during the time it took to collect the coupons and I got my camera.

I used Uncle Willy's lenses to build an epidiascope projector so that I could put on "shows" of the photos that I took of family picnics and outings (Figures 1.6 and 1.7). I built it of wood, painted the inside black, and used two 100-Watt bulbs that heated up to the point that those in attendance experienced some rather strong fumes. To make it a proper show, I hung a white sheet on the wall for a screen. I also wrote and projected commentary cards to accompany the photos that said things like, "Hello Uncle Jack and Aunt Maggie. Many happy returns of the day."

My father also inherited some money, but here Uncle Willy's meanness was catastrophic. To save money, he had not employed an attorney to help him write his will. Instead he had purchased a form for sixpence. The will said that his money was to be divided into six parts, one of which was to a charity. On the strength of this, my father bought the house for which we had previously been paying rent. However, one of the intended recipients of the proceeds of the will had died and the lawyer for the charity disputed the will. The lawyers for the various parties argued about

FIGURE 1.6
A family picnic. From left to right, my brother Jack, his wife Gladys, Joyce with her husband Alfred, and cousin Vera.

FIGURE 1.7
Uncle Jack and my father bringing tea from a local house to the rest of us picnicking at Happy Valley.

this until most of the money was gone, leaving my father a debt that was a great worry to him.

In England, an elementary education was available to all but you were taught very little—mostly how to read and write and to do simple arithmetic. The classes were very large, and you left when you were 14 years old, usually to get a menial job. There was no possibility of escape unless your parents could pay for you to go to a grammar or secondary school, which of course mine couldn't. It was possible to get a scholarship, but there weren't very many of these. The class system based on money was heavily entrenched.

Mr. Spencer, the headmaster of the elementary school I attended, by some means or other, became aware of a poem I had written. To check me out, he told me to sit at a table next to his desk and write poetry. I wrote four poems. After this, he had me stay after school to prepare under his guidance for the scholarship examination.

I remember that during the oral exam I happened to say the word "chimney," which I pronounced "chimley." The examiner asked, "How do you spell chimley?" I spelled it correctly, so he said, "Well why do you say chimley?" I passed, and soon after I went to the new school, which

FIGURE 1.8
Second form at Gravesend County School. I am the first left in the back row.

was called the County School for Boys. I started in the second form, at age ten (Figure 1.8). There was another boy of about my age whose name was also Box (Ronald Box). He was on his bicycle when, unfortunately, he was hit by a truck and killed. The whole school of about 500 boys gathered together in the auditorium each morning with the youngest boys at the back. The headmaster announced from the platform that *I* had been killed. This, I think, was the only time that he said nice things about me. I am told that I walked all the way from the back to just below the platform and said, "Please, Sir, I'm not dead."

My brother Jack and I were among the very few scholarship boys in the school. Although almost all of the students came from families that had more money than mine did, I made friends. But there was one student whose mother would not allow me in the family home when her son and I played together. By contrast, the mother of my friend, "Ginger" Harris, thought I was a good influence on her son. I was never particularly strong, but I made up for my lack of physical prowess by inventing games that the other boys enjoyed.

The second form started immediately with French, Latin, English grammar, English literature, physics, and chemistry. We also began mathematics—first algebra and then geometry—and we began calculus in the upper fourth form. My first math teacher nobody liked. He was sarcastic and unresponsive to questions. But later we had a different math teacher, Mr. Marshall, who for some obscure reason was nicknamed "Banners." He was genuinely anxious that everyone in the class understand the lessons, and he was tireless in explaining difficult points. With his guidance, I quickly moved ahead in the class.

I remember one incident with Banners when one of the boys had brought his pet mouse to school in a little box. During class he was showing his mouse to a friend when it jumped loose and started to run about in the front of the room. Banners pursued it with the pointer, and after a number of near misses, the mouse took refuge under a radiator. The little boy who was owner of the mouse remarked, "Please, Sir: that's my mouse!" Banners replied very apologetically, "Oh I beg your pardon. I didn't know it was a *private* mouse."

Unfortunately, before I could go to the new grammar school, I had been subjected to a "health examination" and it was decided (on what grounds I never understood) that I had to have drops in my eyes for *six months*. As a result, for much of the first year, I couldn't see what was written on the blackboard and had difficulty reading the printed page. I missed such things as the beginning of French, algebra, and English grammar. My father did his best to help and wrote out my homework for me as I dictated. It took a very long time to catch up with my studies, so whereas in elementary school I had been close to the top of my class, at the new school, I had to get used to being close to the bottom. By age 16, when one or two of the boys got to go to a university, I was not among them.

During all this time, Mr. Spencer from the elementary school remained my friend. He was also head of the Sunday school, and our house was on the way to his, so each week after Sunday school, we would walk home together and we discussed just about everything.

One thing I made which was a success was what was called a shocking coil, which was a watered down version of an induction coil. This consisted of a core, which was made of slives of iron wire, and this was wrapped with primary and secondary insulated copper wires.

Each year at our school there was a "prep fair" that raised money for the town hospital. My project for the fair was to make a shocking coil. The coil had two handles, one of which was in a tub of water. The other handle was held by the "client" who was visiting the fair. The client paid sixpence, which was dropped into a locked box. At the bottom of the tub were coins.

The client would reach into the tank to pickup some of the coins, but I had a dial under the table that was connected to a rheostat, and as the customer grasped for the coins, I turned up the dial which gave the person a substantial shock. No one managed to get any of the money until this woman came along, and when it was her turn to reach into the tank, I turned up the dial as usual, but she wasn't the least affected by the shock.

She took all of my coins and she deposited these into the locked hospital box. This left me with no coins with which to lose more clients. Finally Banners, the math master, gave me some money so that I could continue.

I wasn't much good at learning French, but a boy in my class called Newton was, and he wrote a play in French. The French teacher liked it and decided that we should put it on in the school auditorium with parents invited. I had a small part as an "Englishman who didn't know much French," which suited me very well. The French teacher and his wife were very kind, and while rehearsing the play, we had refreshments at their house.

Because my lack of French limited my participation as an actor, I wanted to help all I could in other ways. For example, in the play, someone got shot. I found a realistic-looking toy gun for the actor to flourish, and after much experimentation, I found that a hollow pencil box struck against a plywood panel sounded very much like a pistol shot. I had to watch carefully to synchronize it with what was happening on stage to make it sound genuine.

There was one scene where people were sitting around a table having dinner. To make this look real, I persuaded my brother Jack to ride his bicycle to get six helpings of fish and chips. He arrived a bit early, and the intense smell of the fish and chips was evident for a long time in the theater before and after the actual scene.

Later I had a small part in a *school* play, Shakespeare's *Macbeth*. This was a much more serious affair. My scene occurred at night inside the gates of Macbeth's castle. The audience knows that Duncan, the king, has been murdered, but this is not known to Macduff and Lennox, who are outside the gate knocking to gain admittance. I played the part of the porter who, instead of opening the gate, engages in a long drunken harangue in which he imagines himself porter at the gates of hell. He admits a series of imaginary visitors—a farmer who hanged himself in the expectation of plenty, and so on. Finally he opens the gates to Macduff and Lennox, but the action is further held up when the porter gossips, humorously, with Macduff, a device that effectively increases the tension. At one point Macduff asks the porter, "What three things does drink especially provoke?" to which I replied:

> Marry, sir, nose-painting, sleep, and urine. Leachery, sir, it provokes, and unprovokes; it provokes the desire, but it takes away the performance: therefore, much drink may be said to be an equivocator with leachery: it makes him and it mars him; it sets him on, and it takes him off; it persuades him, and disheartens him; makes him stand to and not stand to; in conclusion, equivocates him in a sleep, and giving him the lie, leaves him.[3]

I was keen on chemistry, and when I left school at 16, I got a job as an assistant to a chemist who managed the sewage treatment plant at Gravesend. I became very interested in the activated sludge process responsible for producing a clean effluent that would not pollute the river, and the first article I ever published was about this topic.[4] While at the plant, my goal was to get an external degree in chemistry from London University. I wasn't paid much, but I was allowed two free afternoons a week to go to Gillingham Technical College where I could attend the necessary courses.

To get to Gillingham, if I had the money, I sometimes took the train, but most often I rode my bike 12 miles along the hilly and busy road

---

[3] William Shakespeare, *The Complete Works of William Shakespeare*, Vol. 2, Garden City, NY: Nelson Doubleday, Inc., nd, p. 799.

[4] Ronald Hicks and G.E. Pelham Box, "Rate of Solution of Air and Rate of Transfer for Sewage Treatment by Activated Sludge Process," *Sewage Purification, Land Drainage, Water and River Engineering*, Vol. 1, June 1939, pp. 271–278. Hicks was my supervisor but did not take part in writing the article.

that passed through Stroud, Chatham, Rochester, and Gillingham. One day my plans almost came to a sudden end when a truck driver's error sent my bicycle and me skidding under his vehicle. His back wheel just missed my head, and my bicycle was a wreck I spent about eight months writing to his insurance company trying to get them to pay to replace it. After months of arguing, they finally did.

To get an external degree in science at London University, you had first to pass the Intermediate Science Exam. After that, with a year or two of further study, you could attempt the Bachelor of Science degree exam itself. I had to go to London to take the intermediate exam, which included a two-day practical exam as well as a week-long written part. My subjects were pure and applied mathematics, physics (heat, light and sound, electricity, and magnetism), and chemistry (organic and inorganic). These were the most difficult exams I ever took, but I passed and they helped me get a grounding in science that has been invaluable ever since.

I believe that it was this basic scientific knowledge that helped me later on to come up with ideas in the development of statistics. It would, I think, be tremendously helpful if, before taking a degree in statistics, there was a requirement to pass a similar preliminary exam in science. A serious mistake has been made in classifying statistics as part of the mathematical sciences. Rather it should be regarded as a catalyst to scientific method itself. Proper preparation for a degree in statistics should be like that for the intermediate science exam described above, which would include running real experiments.

"Contrarywise, if it was so, it might be: and if it were so, it would be: but as it isn't, it ain't. That's logic."

CHAPTER TWO

Army Life

$\mathcal{A}$T home in Gravesend, my family congregated for meals and other activities in the basement kitchen. When it had become clear that war was inevitable, my father, who had experienced the First World War, went to the wood yard and bought a number of very large wooden beams with sections of about 8 inches by 8 inches. With these he fortified the walls and ceiling of the downstairs kitchen so that in an air raid we were less vulnerable.

For many months before war was declared in 1939, people in Britain had been warned what to expect: raids of hundreds of enemy planes dropping bombs and poison gas. Everyone had been issued a gas mask that was carried in a cardboard box. At 11 a.m. on September 3, 1939, Prime Minister Neville Chamberlain announced on the radio that we were at war with Germany. Almost immediately the air raid sirens blew, so my family went to the downstairs kitchen, but once there, we realized that we might be short of drinking water. In the kitchen was our zinc bathtub in which each family member took a weekly bath. We hoisted the tub onto the sink and filled it with water from the tap, forgetting how heavy it would become. In our attempt to lift it down, most of the water spilled onto the floor, so apart from everything else, we had a small flood. After about three quarters of an hour, we heard the "all clear" siren. Our first air raid warning had in fact been a false alarm. This was the beginning of the "Phoney War," a period of about

An Accidental Statistician: The Life and Memories of George E.P. Box, First Edition. George E.P. Box.
© 2013 John Wiley & Sons, Inc. Published 2013 by John Wiley & Sons, Inc.

eight months when neither side did very much, at least in terms of a land war.[1]

As a teenager in the 1930s, I was interested in politics. I was particularly angry at the British government because it seemed they had done nothing to stop Adolph Hitler. His aggressions and absorption of one country after another clearly pointed to a plan for world domination. Six weeks after the war began, I turned 20, the age required at that time for enlistment into the Army. I stopped working on my chemistry degree, went to the nearest enlistment office, which was in Chatham, and joined up.

During my first week in the Army, I was told to read Company Orders each day and to perform any duties required of me. Sure enough, on a long list outside the Company Office, I found my name, and next to it were the words "key man." I asked somebody what I was supposed to do as the key man, and they explained that at some time during the evening, the bugler would sound the "Buckshee Fire Call." (This was the fire call followed by two Gs, which meant it was not for a real fire.) When you heard this, you ran down to the Company Office, and when they called your name, you shouted "present." So I did that, but then I wondered what I was supposed to do *as the key man*. I asked a junior NCO.[2] He said, "When you hear the Buckshee Fire Call you run down to the Company Office and answer your name." "But," I persisted," What am I supposed to *do*?" (After all, "key man" sounded rather special.) He said, "I've already told you. When you hear the Buckshee Fire Call, you answer your name at the Company Office."

I made myself a nuisance, and my enquiry eventually got as far as the Sergeant Major who went through the same routine. So I said, "But in a real emergency what does the key man do?" This stumped him completely. He asked me to come and see him tomorrow. By this time I was a marked man—a raw recruit making all this trouble! But after a great deal of research, they discovered that the "key" was a spigot that was used to turn on the water in case of a fire. So I asked where it was. Again this caused much consternation, and for a long time, they couldn't find it.

---

[1] The war at sea was a different story. It began with the torpedoing of the British liner, SS Athenia, soon after Britain and France declared war on Germany. We suffered huge losses of our merchant ships. We had to import a great deal of our food, and the enemy's intention was to starve us out.
[2] Noncommisioned officer.

Fortunately, there never was a real fire while I was the key man, but I had learned an important lesson about how things "worked" in the Army.

At the beginning of the war, my company was busy erecting barrack huts. These came in prefabricated sections—walls, roofs, doors, windows, stoves, and so on.[3] We had to go some distance in our trucks to collect these sections from a different branch of the army called the Service Corps. When we weren't building barrack huts, we were frequently on guard duty. This meant lots of "spit and polish" to pass inspection, and staying up all night. It was winter, and the "guard room" consisted of a tent that, especially when it rained or snowed, was cold and miserable. It occurred to us that we might just as well *steal* the necessary prefabricated sections and build *ourselves* a guard room with a roof and a stove. One dark night we did just that.

I am quite sure that our commanding officer, a very cheerful major, was aware that a new building had suddenly popped up; guard rooms didn't appear overnight. I'm even fairly sure that he *approved* of our initiative, but that, of course, was unofficial. Meanwhile the Service Corps sergeant had noticed that some prefabricated sections were unaccounted for. He was suspicious and therefore loath to give us any further supplies, so my little group couldn't get on with our work. I was a only a lance corporal, so to outrank the Service Corps sergeant, I went to my commissioned officer who was a brand new second lieutenant even younger than me. I told him about our difficulties with the sergeant, and he sighed and said, "Yes, I know. He won't even *speak* to me."

When I first joined the army, there were all kinds of guys from different parts of Britain. They had very different backgrounds and accents, but we got along well together. In the next bed to mine, there was a coal miner. He seemed to be having a lot of trouble with a letter he was writing. When he showed it to me, I realized that someone was trying to extract money from his wife for some furniture she had bought on an installment plan. I knew that Parliament had passed a law stating that you could not do this to people serving in the armed forces. So I wrote a letter for him spelling this out and he was very grateful. A few weeks later, we all had to dig a slit trench six feet long by six feet deep. I

---

[3]The huts were wide enough to accommodate two beds end to end with a walking space in between. You could make them as long as you liked by adding more sections. The stoves were upright with a pipe that went through the roof.

was not very strong and hadn't got very far when he had already finished. He came over and said, "Get out of there!" and he quickly dug my trench for me.

Slit trenches would plague me in other ways. We had dug a number of these near the barracks and the sergeants' mess. On one occasion, it had been raining for a number of days, but finally it had cleared. Later in the evening, I thought it might be pleasant to take a walk outside with a good-looking young lady. It was pitch black out, and we hadn't gotten very far with our walk when we both fell into the same trench, which was full of water. We both got soaked, and explanations were difficult.

I've forgotten our Company Sergeant Major's name, but we called him "The Bull." Everyone was afraid of him, but I think he had some admirable traits. In particular, he always wanted to be absolutely sure that he was familiar with everything that the Company needed to do. People were frequently sent to courses. If there were a new type of trench mortar, for example, a sergeant might be sent on a course of training and would return to instruct our company on how it was operated and how you could take it apart and put it together again. Or there might be a new anti-tank mine: How, for example, was this to be set? Without delay, the Bull required anyone returning from such a course to teach him the new techniques, and while being instructed, he would forget all about rank. He would smartly carry out instructions until he could do the task as well as his instructor. He felt he must know all the things that everybody in the Company knew and be able to complete a drill better than anybody else.

When the Bull asked you a question, he expected a quick and plausible answer, and if he got this, then he was satisfied. For example, for a time we were stationed at a beautiful seaside resort on the south coast of England, partially evacuated at the time because of the possibility of invasion. One day there wasn't much to do so three of us goofed off and went for a swim. The senior of our small gang was a Corporal Cornford, so it was to him that the Bull, who unexpectedly appeared, addressed his remarks. "What are you *on*, Corporal?" he demanded. With great presence of mind, Cornford replied, "Teaching Lance Corporal Box to swim, Sir." I think the Bull knew as well as we did that this wasn't true, but because it was said smartly and without hesitation, he accepted it and gave us a detailed lecture then and there on the best way to teach a beginner to swim.

There was a great hill that dominated the resort town where the company was stationed. Our commanding officer must have been at a dinner party with the local bigwigs one night because the following morning he got us all together to tell us about an iron gun that was on top of the hill that the city fathers wanted removed. It must have been 200 years old and was very large, close to 20 feet long. He asked us all to write a reconnaissance report on how it might be removed. Very quickly it became obvious that we didn't have appropriate equipment to move such a heavy object. All we had were explosives. I had a scheme to first blow a very large hole into the ground and then roll the gun into it with explosives attached to it. We would then fill in the hole and blow up the gun. My suggestion was not adopted.

Someone claimed that they could put explosives on the gun as it sat on the hill and so finely calculate the charge that the resulting explosion would merely crack the gun. The gun could then be broken into parts and hauled away. When we did this, however, our calculations must have been a bit off. There was an enormous explosion and bits of gun fell from the sky all over the town, and greatly surprised some fishermen who were out to sea in their boat. A lady with a very proper accent stated emphatically that her piano had been shot completely through with *little holes* from the falling debris. By some miracle, there were no casualties, but for some days, we went about the town mending things.

The Company consisted of platoons, each divided into four sections of 10 to 12 men. In the summer, we had lectures during which we sat on the ground in an open field and an NCO lectured about some aspect of our duties. It might be about the names of parts of a machine gun and how these parts could be assembled or any number of other things. One morning we had for our instructor a particular sergeant who was what we called an "old sweat"—that is, a "lifer" in the military. He knew very little except for one topic that he knew very well and that was "knots and lashings." One day I remember he was making a hash of some topic or other when in the middle of a sentence, he abruptly switched to talking about how to tie a bowline. When we looked around, we saw that he had seen the colonel who had come over to listen. As the sergeant was well aware, knots and lashings were also the colonel's favorite topic. "Very good sergeant," the colonel remarked approvingly. He then recounted *his* way of remembering how to tie the knot. All I recall is that "the rabbit ran around the tree." There was a saying in the Army: "Bullshit baffles brains" and it was very true.

The commanding officer for our company worked very hard at getting us properly trained. In particular he taught us how to avoid "booby traps." Once, for example, we were working in the trees putting up camouflage nets when along came a truck that we thought was the tea wagon bringing us a well-earned cup of tea and a bun. We left off working and gathered around the truck in expectation. Then out jumped four men with submachine guns trained on us. We did not need to be told that we would all have been dead had this been the enemy. He and his crew also kept a supply of "thunder flashes," which made a tremendous flash and a deafening crack. In the middle of the night, if you were a member of the guard and he suspected that you were not sufficiently alert, you could expect a rude awakening when one of these was thrown at you.

We used to go on exercises that usually lasted three nights. We called them "stunts." The company, in 30 or so 1500-weight trucks, would drive onto Salisbury Plain to follow a route described in the operation order. When (and if) we got where we were supposed to be going, we carried out a mission, such as a demolition, and found our way home. We had a detailed ordinance map showing the proposed route, but we moved only at night and the only light you were allowed was a very faint one underneath the truck in front of you. We frequently got lost, finding ourselves in a farm yard with everybody swearing and the Sergeant saying, "Well the map must be wrong." Even worse, when it rained, one or more of our trucks might get bogged down and had to be dug out.

On one occasion, I had just returned from leave and found our vehicles all lined up to go on a stunt. I was too late for the briefing, but being a lance corporal, I put on my equipment and took my place in the front of the truck. I asked my driver, "What are we supposed to be doing tonight?" He said, "I don't know Corporal, but I'll tell you one thing I'm absolutely certain of: it's bound to be a balls up."[4]

Strangely enough, during the war, I was never in any danger except when I was on leave! Salsbury was far from the bombing, but Gravesend was on the direct route of the bombers going to London. As it turned out, our house was never hit, but a number of neighbors' houses were.

---

[4] Known in the U.S. Army as a SNAFU.

The fact that a house was bombed did not mean necessarily, or even usually, that people were killed. When people were "bombed out" after a raid, the neighbors rescued them from the ruins, gave them tea, fed them, provided shelter, and helped in every way they could. The war seemed to bring out the best in people. After it was over, I heard it said many times, "Why can't people be like they were during the war?"

Everything was rationed or, if not rationed, very hard to get. On one occasion, when I was returning from leave, I saw a queue outside a shop, so I joined in and asked what we were queuing for. It was for alarm clocks, so I waited my turn and bought one. I set the clock to the correct time and put it back in its cardboard box and put the box in my kit bag.

On the train, I sat opposite an old lady. After we had traveled some distance, she asked if I could tell her the time. I said, "Yes, certainly," and standing, I removed my kit bag from the overhead rack, fiddled about inside it, eventually retrieved the cardboard box, opened it, and looked at the clock. "It's ten minutes to three," I told her. Then I put the clock back in the box, put the box back in the kit bag, and put the bag back up on the rack. My companion appeared a little dazed.

England had imported about two thirds of its food before the war, and these supplies were cut off by the Germans once war broke out. The rationing that began in January 1940 was severe, and the British Army was included in this, although soldiers fared somewhat better than citizens. Among the most sought after food items that were on the restricted list were sources of protein. Weekly rations per person consisted of one egg, three pints of milk, and a very small portion of meat, which was rationed according to its value.[5]

The American soldiers who were in England during the war had abundant food, as far as we were concerned. Before coming to England, all U.S. GIs had been issued a small book advising them of British customs and of how the war had affected England since 1939. One passage read, "If you are invited to eat with a family, don't eat too much. Otherwise you may eat up their weekly rations."[6]

One year I became friendly with a GI and we used to sometimes meet at a pub in Salisbury. When Christmas came, he invited me to the

---

[5] Weekly meat rations were limited to the value of one shilling six pence—or 6p in today's terms.
[6] *Instructions for American Servicemen in Britain 1942*, United States War Department, 1942.

Christmas party that the American soldiers were having. When I went to the gathering, I was simply amazed: They had everything that you could possibly want and anything that could not be bought in England. So for that one day, I lived in Arcady.

Food scarcity, and especially meat shortages, continued long after the war ended. Things were so bad that by the early 1950s, it became legal to sell sausage that had very little meat content at all. One member of the House of Commons famously asked, "Would the right honorable gentleman be prepared to say at what point a sausage becomes a cream bun?"

I sometimes like to play the piano. I'm not very good, but how I arrived at what I *can* do is of some interest. When I was about eight years old, I was made to take piano lessons from my cousin. That involved a two-mile walk to her house and a two-mile walk back. I never really made any progress because Vera taught me piano as if I were learning to type, but my mind was, and is, such that I can never learn anything unless I know why.

After I had taken the intermediate science exams, and before the war broke out, I was looking for something to do in my new-found free time, and I took up the guitar, using the one that Uncle Willy had bequeathed us. My instructor was from the merchant navy. (He was killed early in the war, as were so many in that service.) He explained to me about chords and chord progressions and I became quite interested in this, and particularly in composers like George Gershwin and Cole Porter who used clever chord sequences. Later, when I started tentatively playing the piano, I transferred the guitar chords to the piano notes, and this at last made sense to me.

When I was stationed in Salisbury, I asked my father to send my guitar to me. I went down to the railway station each day, and sure enough, it eventually turned up. He had built a guitar case that looked a bit like a small coffin, but it worked. There were a number of us who were interested in playing jazz. In addition to a trumpet, a trombone, drums, and my guitar, there was a pianist named Smudger Smith. He was quite small and could play the piano in almost any key, drunk or sober. We got some music and started playing together.

Smudger Smith had a friend who looked like a heavyweight boxer. This friend was not very bright and did not play an instrument himself, but he believed that his duty was to supply Smudger with plenty of beer. So while we played, he would make sure that at any given time, there were two pint glasses on top of the piano appropriately filled.

After a bit we got hired to play at dances. All went well until one evening we were to play at a temperance society dance. Smudger's friend could not, of course, find any beer at the dance, so he walked up the road and bought two pints at the nearest pub. When he returned with the beer, the men at the door tried to stop him, but he elbowed them away and delivered his goods atop the piano in the usual way. As I recall the supply was regularly replenished during the session—an unusual temperance meeting!

On my 21st birthday, I was sharing a barrack room with seven other guys. One of them and I felt the need to celebrate my birthday, so we caught the bus into Salisbury and visited a few pubs. The beer in some was better than in others. Before catching the bus home, we became somewhat morbid thinking of our friends in the barrack room who didn't have any beer, so we bought a good supply to take back with us. Fortunately the large pockets in our great coats accommodated several bottles. When we got back, we announced to our colleagues that we had brought them beer, but when we emptied our pockets, we were surprised to see that the bottles were drained. In our celebratory state, we had forgotten that the bus had been delayed and that we had drunk them on the way home.

One morning I was a bit late on parade, and when I got there, I found we had been told to form a single line. An NCO divided the line into two halves. The front half was marched off to where we did not know. The half I was in remained. I discovered later that the first half had been shipped off to Singapore, which fell to the Japanese on February 15, 1942. All had become prisoners of war, and most of them were never heard from again.

Early on in the army, I trained with the engineers and, in particular, learned how to demolish bridges. You don't actually blow up a bridge; you use explosives to cut the girders that are holding it up. There are simple formulas that tell you how much explosive to use, and it was because I understood these that I got to calculate the charges and work the exploder in our practice demolitions. There was no doubt that I was better at demolitions than I was at knots and lashings.

FIGURE 2.1
Porton, Colonel Collumbine center. I am to his left.

The Experimental Station at Porton Down

Before I could employ my demolition skills in real warfare, someone in the Army found out that I had a background in chemistry, and I received orders to report to the Chemical Defence Experimental Station at Porton Down. This was near Salisbury, in the south of England. It was fully expected that the Germans would eventually use poison gas, and the purpose of the experimental station was to find out what to do about it. Some of England's best scientists were there. My boss, Dr. Harry Cullumbine, for example, was a professor of physiology dressed up as a colonel. I became a lab assistant dressed up as a staff sergeant (Figure 2.1).

The army decided that every army unit should be issued a small sample of liquid mustard gas so that volunteers could have a drop put on their skin and see how a painful blister appeared. In army operation orders, you always had to sign off on a line that said, "All Informed," but I'm afraid this protocol was not followed with mustard gas samples. Cullumbine, who was the expert on the treatment of mustard gas casualties, was kept very busy by the misuse of these samples. In one instance, for example, someone must have decided that the black liquid in the can was for painting the stove. The result was close to disastrous for the people who slept in that barrack room.

Initially my job was to make biochemical determinations in experiments on small animals. The results I was getting were very variable, and I told Cullumbine that what we needed was a statistician to analyze our data. He said, "Yes, but we can't get one. What do you know about it?" I told him I had once tried to read a book about it by someone called R.A. Fisher, but I hadn't understood it. He said, "Well *you* read the book so you'd better do it." So I said, "Yes Sir."

I wrote to the Army Education Corps, which sent me a list of books on statistics that I carefully studied.[7] I soon realized that it was not just a question of statistical *analysis*, but also that we needed to carefully *design* our experiments using statistical principles. Before long I was given some assistants, and for the rest of the war, I spent my time planning, supervising, and helping to run experiments both in the lab and outside on the range where we simulated warfare. Thus, I changed my plan from becoming a chemist to becoming a statistician. For the rest of the war, I was in fact the only one at the station responsible for statistics, and Cullumbine and I wrote several articles describing the results of our experiments.[8]

To illustrate the power of experimental design, I show here a simple experimental arrangement that we used in the lab to find the best

---

[7] Two of the books were Fisher's *Statistical Methods for Experimenters* and his *Design of Experiments*. The others were about applications by followers of Fisher. I remember *Statistical Methods in Forestry and Range Management*, which had an excellent account of the "method of Least Squares." Another was about using statistical design to improve teaching methods.

[8] H. Cullumbine and G.E.P. Box, "Treatment of Lewisite Shock with Sodium Salt Solutions," *British Medical Journal*, April 20, 1946, pp. 607–608; G.E.P. Box and H. Cullumbine, "The Relationship between Survival Time and Dosage with Certain Toxic Agents," *British Journal of Pharmacology*, Vol. 2, 1947, pp. 27–37.

treatment for mustard gas blisters on soldier-volunteers. A very small drop of mustard gas liquid applied to the arm of a volunteer caused a blister about three quarters of an inch in diameter. It was similar to that produced by an ordinary burn, but it was more difficult to treat and took longer to heal. There were difficulties in making comparisons between different treatments because the ability of the body to heal is different for different people, and for different parts of the body. There was a further problem in that we needed to get valid results quickly.

The response measure that we used was healing time in days. For each experiment, we had six volunteers (labeled 1–6 in the arrangement), and six drops of mustard gas were placed at six different places on each arm of each volunteer (labeled A–F). There also were six different treatments (a–f) allocated as indicated in the table. You will see that each treatment was tested once on each of the six volunteers at each of the six positions on the arms. So each treatment occurred once with each volunteer and once in each position. Thus, differences among the volunteers and differences resulting from positions on the arm could be calculated and eliminated. This arrangement, which Fisher invented, is called a "Latin Square" design. The use of similar but sometimes much more complicated arrangements is part of the general science of "statistical experimental design."

				Volunteers				
			1	2	3	4	5	6
Left Arm	{	A	a	b	c	d	e	f
		B	b	a	e	f	c	d
	{	C	c	f	a	b	d	e
Position on Arm	{	D	d	e	b	a	f	c
Right Arm	{	E	e	d	f	c	b	a
	{	F	f	c	d	e	a	b

Treatment a,b,c,d,e,f

There are many ways in which you can rearrange the letters in this diagram and still get a Latin Square, so Fisher enumerated all possible rearrangements and proposed that the experimenter pick one at random. He also pointed out that then, without making any assumptions about probability distributions, you can make a statistical significance test to reveal what treatments are best by considering what actually happened in relation to what *might have* happened with all other possible random arrangements of the same data. Later, tests of this kind were rediscovered by mathematical statisticians who called them "nonparametric" tests.

One day at the lab, I was having trouble with a particular statistical problem, and a senior medical scientist there suggested that I write to R. A. Fisher about it.[9] I thought Fisher would be much too busy to talk to me, but he replied, asking me to come to see him and to bring my data. The Army, however, did not have a procedure to send a sergeant to see a professor at Cambridge, so they made out a railway warrant that said I was taking a horse there.

When I arrived at Fisher's house, it was a beautiful day. He said, "Let's go and sit under that tree in the orchard. I'll look up the probits and you look up the reciprocals, and we'll plot the data." I didn't know that that was what we were going to do, but in doing it, my problem was quickly solved. Fisher was extremely kind and spent the whole day with me.

There were a number of characters at the Porton Down experimental lab. Down the corridor where I worked, for example, was a pathologist who, like many others, had been put into an army uniform for the duration of the war. I will call him Major Long. Major Long unwittingly provided humor where humor was much needed. For example, there was a rule in the army that all "other ranks" must salute officers and that officers must return the salute. Major Long rode a bicycle to work, and this made him very uncertain about which hand he should salute with. We enjoyed saluting him to see whether he would fall off his bicycle.

One day my colleagues and I were working in the lab, and Major Long's head appeared around the door. He said, "Have you seen a rabbit?" We told him, "No." After a while, he reappeared and said, "It had a *tube* in its mouth."

On another occasion, Major Long came to see me about the layout of an extensive field trial. I had arranged that the 40 sampling points were

---

[9]The reason that the scientist at the station knew Fisher was that Fisher did not get on very well with Karl Pearson and his people at University College, where they both had laboratories, and rather than eat with *them*, he crossed the street and had lunch at University College Hospital.

randomly distributed over a particular area. He came to me and said, "They can't *possibly* be random." So I showed him a map on which I had divided the entire area into 100 numbered squares, and I explained that we took 40 items from a table of random numbers to determine which squares became our 40 sampling points. He looked at me suspiciously and said, "Ah yes! But what about the mirror image?" He was an innocent soul. One day he was tired and explained to us that he had been up most of the night teaching his wife biology.

An outstanding scientist whom I came to know was Britain's leading pharmacologist, Professor John Gaddum (later Sir John). He was himself a competent statistician, and we had a number of discussions about the work that was going on. He was unimpressed with self-important figures and unintelligent generals. At the beginning of the war, they wanted to dress him up as a colonel and keep him at the station. He would not go along with this, however, and returned to his university in Edinburgh, telling them that they could find him there if needed. He was particularly interested in an experiment that I had designed for an American researcher concerning lewisite, a chemical warfare agent. If you were unlucky enough to get a tiny drop of lewisite in your eye, after a short interval, you were blinded. The investigator was trying to find out how to prevent this, using the eyes of rabbits for experimentation.

The problem was that although the two eyes of a single rabbit were comparable, comparisons between the eyes of different rabbits were not. I was rather proud of a complicated statistical design that I developed in which all the important treatments could be investigated with no more variation than that between the eyes of single rabbits. My investigator asked me to write an appendix that explained this complex design. After the report was published, Gaddum came to see me and said he'd read the report but what had happened to my appendix? I explained that the people in what was called (rather appropriately) the "main block" had deleted it, presumably because it was written by a mere sergeant. (These "in-charge" people were not scientists, of course, but high-ranking civil servants.) Gaddum was incensed, and he took me by the hand and said, "Come with me." We marched up to see the top brass. He asked them on what grounds they had removed my contribution, and getting no satisfactory reply, he became quite angry. He banged on the table and said, "Put the damn thing back!" They did.

During the war, the British Army had an outfit called ENSA that was supposed to entertain the troops. Each week they sent us singers, dancers, magicians, and so on, all second rate and uniformly awful. I was so disgusted with one lot that in an unguarded moment in the sergeant's mess, I announced that I could put on a better show myself. Some scoffed at this idea, so to prove them wrong, I got together with like-minded people to make my point.

Obviously one thing we needed was a chorus of girls. At the station, there were women from the Army (ATS), Navy (WRENS), and Air Force (WAAFS). The WAAFS had the best legs, but they refused to be in the show if we allowed girls from the other services to be in it. I argued with them, but in the end, they prevailed and the chorus was all WAAFS (Figure 2.2).

The show was called, "You've Had It," and we wrote a song for a curtain raiser:

We're a new show
Called "You've Had It"
Not a blue show

FIGURE 2.2
We put on a show.

And we're glad it's
Filled with laughter and with music
Hope you like us in our new show.

We had various "acts." In one of these, the curtain went up revealing a folding screen on center stage. One by one female garments—jacket, skirt, bra, panties, and so on—were hung up by someone on the other side of the screen. Then the screen fell down to show a fully dressed young lady ironing her clothes.

In another act, the curtain went up to reveal a white sheet illuminated from behind. The shadow on the screen showed an inert figure lying on a table. The audience was told that this was the sergeant major who was about to undergo an operation. The shadow of the "surgeon" appeared carrying what appeared to be a huge saw. The rhythmic screeching of the saw blade was accompanied by loud howling. Then various objects were drawn out, apparently from the sergeant major's stomach. Among these was a dead rat.

Our show was a hit. It's not quite true, by the way, that the ENSA shows were all awful. Once, we had Glenn Miller who was, of course, first class. I remember they played "When We Begin to Clean the Latrine" to the tune of "Begin the Beguine."

In another homegrown effort to entertain the troops, I took part in a production of the play, *Cinderella*, organized by a Major Kirwin. I played Cinderella, and the glass slipper was played by my Army boot. Our play deviated in many other ways from the original story.

The show went on for two successive nights. On the first night, there was a bit of a prop malfunction. Cinderella's coach was pulled by a pretend horse powered by two lance corporals, one for the back legs and one for the front. At some point, suddenly, much too early and quite unexpectedly, the horse and coach appeared on stage. This stopped the show because no one knew what to do. Kirwin was incensed. He ran across the stage and tried to force the horse off, but what he succeeded in doing was pulling its head off. This exposed the top half of a lance corporal who blinked pathetically into the bright lights. Kirwin didn't help matters by saying something like, "I've seen some idiots in my time, but you're the worst."

On the second night, things went better, but there was a colonel from the artillery in the audience who had seen both performances. He was not particularly bright, and after the performance, he was heard to say to Major Kirwin, "I *did* enjoy the show the second time, but you left out the best part, you know—where you pull the head off the horse."

Another mishap occurred during the scene when Cinderella was supposed to go miraculously from rags to riches. The lights went out, and I was required to do a very quick change. Jessie, my wife to be, was in the wings to help me with this. Unfortunately, I was in such a hurry that I brought up my knee and accidentally gave her a black eye.

Jessie and I had met in the Army. She was a sergeant in the ATS and worked as a secretary for an officer training group stationed near our experimental station. She was a great companion, and we enjoyed long walks on Salisbury Plain, found places to eat in the evening, and shared books together. We were married at a church in Cheshire in 1945. The war was winding down, and most soldiers would soon be home, but my tour of duty in the Army was not over.

The poison gases that we had been studying in England were mostly those that had been used in the First World War (WWI), and there was little that was new. But near the end of the war, the Allies discovered shells in Germany containing a new group of toxic nerve gasses (tabun, sarin, and soman). I was there when we received a small sample of one of these at our experimental station. Three of us—Professor Gaddum, his assistant Mac, and me—were there when we witnessed some preliminary estimates of the toxicity.(There were ways of testing highly poisonous substances without being exposed to them.) Mac prepared a highly diluted sample and injected a rabbit. Even at this minute dose, the rabbit died at once. Gaddum was surprised and asked Mac to check the dilution and to do it again, and again the rabbit died immediately. So Mac tried a tenth of the dose, and then a hundredth, with the same result, and it was soon clear that we were dealing with substances that were orders of magnitude more toxic than anything we had known. When vaporized, they produced gases that had an immediate effect on the central nervous system, quickly resulting in death.

In short order, Britain sent a group of experts and their assistants on a secret mission to Raubkammer, the enemy's experimental station at Munster Lager, in Northern Germany (Figure 2.3). I was part of

Tr. Üb. Pl. Raubkammer bei Munster=L.   Stabsgebäude

FIGURE 2.3
The German Experimental Station at Raubkammer.

this team, and I helped to design and analyze some of the field trials necessary for the study of these frightening new substances. We crossed the Channel on tank transporters with about 40 trucks filled with lab equipment. Sitting next to the drivers were various kinds of experts in chemical warfare wearing uniforms from the Army, Navy, and Air Force as well as some civilians. We slowly made our way up through the destruction. We must have seemed a strange lot as we drove through the ruins of Belgium and Germany. I remember coming to an intersection where army police were directing lines of tanks and other army vehicles. They looked at our peculiar convoy in amazement and said, "What's this lot then?" We told them that our mission was very secret, and finally we produced a document that satisfied them and they allowed us to proceed.

Upon arriving at Raubkammer, we saw how the Germans had cleverly camouflaged their research station. For instance, the chemistry lab and the physics lab looked like farmhouses that were situated a considerable distance apart. To make it easier to communicate, they had developed small electric cars that they recharged every night. We made good use of these.

The Germans had also developed a superior system to study gas shells. At our experimental station in England, to simulate reality, the artillery had fired gas shells onto the range from a couple of miles away. The firing wasn't very accurate and often landed uncomfortably close, prompting us to run for cover. At the German station, however, there was a tower close to the layout with a gun on top that used a reduced charge so that the speed of the shell and the angle at which it landed simulated firing from a long way away. While I was there, the man who performed this operation was called "explosives worker Hellman," and although he had had the same job under the Nazis, no one found this arrangement uncomfortable. I had taken some German in school and tried to speak it when I first got to Raubkammer, until a German worker at the station said, "I tink it vud be bettah if we spoke eenEenglish." I took the hint.

It was easy to forget how toxic these nerve gasses were. We were used to field trials using WWI gases such as phosgene and lewisite when all you had to do was take a sniff and you would have plenty of time to put on your respirator. That would have been fatal with these new gases. Once, for example, I entered the lab at Raubkammer and thought there had been a failure in the electricity supply because it seemed to me that all the lights were dim. What had happened, actually, was that a minute dose of nerve gas had caused my pupils to constrict.

From time to time, we needed small items from the village down the road. On these trips, a sergeant from the German Army acted as my chauffeur. I'll call him Sergeant Shultz. He was a big man, and we communicated very well, mostly with signs. Although his uniform had been mended so many times it was a truly pitiful sight, he kept his dignity. I imagine he had seen quite a bit of the fighting, perhaps on the Russian front.

Down the road, there had been a prisoner of war camp for Italian officers. They had changed sides sometime previously, and they wandered around in their fancy uniforms waiting for repatriation to Italy. Shultz respected British troops, but he regarded these Italians as rank traitors, and when we saw them walking along the road, he became uncontrollable. He stuck his head out of the window of the car and, while making obscene gestures, shouted, "Macaroni! Macaroni!" as we passed. I saw his point of view.

There was also a number of standard cars available to us, and from time to time, we used them to visit cities such as Hamburg and Hanover. I think that most people in England and the United States have no idea of the intensity of devastation from aerial bombing that Germany had experienced in the second half of WWII. The damage was scarcely credible. There were huge areas in which there were not even ruins; everything was completely flat so far as the eye could see.

The feeling against the RAF was intense among the Germans. At Munster Lager, we received our meals in the same facility that had been used by the previous German investigators with the same people running the kitchen and handing out the food. I was going through the line with a colleague when the young lady behind the counter recognized my friend's RAF uniform and exclaiming, "Luft," threw a bowl of soup all over him.

When we had first arrived at Munster Lager, we had taken over the 200 or 300 Germans that had run the show—from cooks to chromatographers— and when we were leaving, they were very sad. How were they to live now? The best we could do was to give them our remaining supply of absolute alcohol. They were happily singing when we left.

Being at the German research station delayed my discharge from the Army. V-E Day was on May 8, 1945, but I remained in Germany for six months, until the end of 1945. When I finally did return to England, the Army gave me a medal and I was "demobbed"—that is, demobilized. By the time I got to it, the process of demobilization had become very efficient. The demobilization office was close to Exeter, in the west of England. You went on the train as a soldier and came out on another train a civilian. At the demob station, we first received some new underclothes, then went on to a larger room in which suits of various colors and sizes were available, along with shirts and ties and shoes. My recollection is that we came out looking quite smart. It was strange to see the Army being so efficient, but I was one of the last to be "demobbed," so they had had a lot of practice.

In England, there was an arrangement similar to the GI Bill in the United States. The government paid for me to attend University College in London for three years to study statistics under Professor E.S. Pearson.

But getting there was not as easy as it sounds. The Army said they would release me if I could show that I had been accepted at the University. The University said they would accept me if I could show that I had been discharged from the Army. Most of the time I am a truthful person, but when you are dealing with bureaucracy, careful lying is sometimes essential. I forget whom I lied to (I expect it was the Army—they were used to it), but I did get my discharge.

While I was at University College, Jessie and I lodged with a man who was working on his Ph.D. Jessie acted as his housekeeper and cook in exchange for our rented room. Later we lived with my sister Joyce and her husband Alfred. Jessie and Joyce were experts in making our skimpy rations go a long way. Alfred used to cycle home from work at midday, which left him little time to eat his meal, but he refused an offer from Joyce to pack him a lunch each day. However, when Alfred's brother Barrie, who had undue influence over Alfred, said, "Alfred, instead of cycling back and forth, why don't you ask Joyce to pack you a lunch to take with you?" and Alfred *did* ask Joyce to pack him a lunch: she became quite annoyed.

Although living conditions were difficult after the war, we also had some happy times together. On a number of occasions we managed week-long houseboat trips on the River Thames with Joyce and Alfred and Jessie's parents (Figure 2.4). Once it gets past London, the Thames becomes very different. It is much narrower, beautifully clean, and passes through lovely countryside. It wanders about and eventually passes through Oxford. Fully equipped houseboats were available for hire, and it was pleasant in the summer to rent one and to explore the river leisurely.

While I was at University College, there was a day set aside for visiting relatives. My mother came, and I showed her the Statistics Department. Next door to it was the Genetics Department. There I showed her some Ishihara charts that have colored dots arranged so that if you had normal vision, you would see one number, and if you were red-green colorblind, you would see another. I showed one to my mother and said, "If you were colorblind, you would see a six." She said, "Well it *is* a six." Dr. Kalmus, the professor of genetics, happened to be passing by and asked, "Is someone colorblind?" I replied, "Yes, my mother." "I don't think so," he said. Of course it turned out that *I* was colorblind. The

FIGURE 2.4
Boat with Joyce and family.

professor explained that only 0.4% of women are colorblind, as opposed to 7% of men. At this point, I was 26 years old, had been in the Army, and through the war without a clue about this.

My friend Merve Muller recently reminded me that while we were both at Princeton, he was a passenger in my car when I slowed unexpectedly at a traffic light. I said, "I know that when the lights are vertical, the top one is green. But *these* lights are horizontal. Is the one on the left green?" He looked surprised at my question, and feared, perhaps, for his life, so I told him of my colorblindness.

Merve also recently recalled a time when we were both at the same conference in Texas. I showed him some very nice gray shirts that I had purchased on sale in the hotel and urged him to take advantage of the special price. Merve did not have the heart to tell me that my new shirts were in fact a vivid shade of pink.

University College was at one end of Gower Street and further down was the London School of Tropical Medicine where the Royal Statistical Society had its meetings. I only had to walk a few hundred yards down Gower Street to go to these meetings. In particular, there were four "read" papers presented each year. These special research papers were printed in advance and circulated to the membership before the meeting. Because we had already read the paper, only a brief introduction was made by the author at the meeting, and most of the time could be devoted to discussion. By tradition, the "proposer of the vote of thanks" said what he thought was good about the paper and the seconder said what he thought was not so good. After this, there was a general discussion by the fellows of the Society and often a number of famous statisticians spoke.

On one occasion during the meeting, an idea occurred to me that no one else seemed to have thought about. So even though I was only a student, I put up my hand, was acknowledged by the chairman, and made my point in about three minutes with the help of a piece of chalk and a blackboard. Immediately after the meeting was over, a stranger came over to me and said, "I'm George Barnard. What are you doing this evening?" When I replied, "Nothing," he said, "Well let's go and get a bottle of wine and have dinner together" (Figure 2.5).

George, it turned out, was a professor at Imperial College. From that date forward, he became my friend and mentor and there began a close association that lasted until the end of George's life. For example, it wasn't long after I met George that I became confused about the theory of least squares, an important statistical technique for estimating unknown constants. I was dissatisfied with a "proof" I had been taught at University College. It seemed to me that it was not rigorous and only covered a special case. I asked George, and he said, "I'll show you how Gauss did it." He wrote three lines of matrix algebra and that was it. It was beautiful and completely general.[10] In a subsequent class, there was a complex problem in an exam that took almost all the students a long time to do, but using the Gauss proof that George had shown me, I completed the problem in five minutes.

---

[10]The Gauss proof showed that the squared differences between the least-square estimates and *any* other set of linear estimates was always greater than zero!

FIGURE 2.5
George Barnard.

It took me 18 months to get my undergraduate degree in statistics with first-class honors. I spent the rest of the three years for which I was supported doing graduate work. I think there were seven students in the class the first year of instruction. Of course much had happened to all of us during World War II. Unfortunately, we had an instructor who didn't seem to notice this, and persisted in treating us like school children. She would point a finger at one of us and demand, for example, that he or she "Define an expectation." This nonsense was put a stop to by a quiet student called Westgarth. He was an ex-major in the tank corps who had been a commander in the desert in North Africa fighting against General Rommel. She pointed to him and said, "What is a random variable?" Eyeing the instructor fixedly, Westgarth slowly put his feet on the table, leaned back, and said, "I haven't the *slightest* idea." That seemed to cure her.

Thankfully, the poor classroom technique of this particular instructor was the exception. I was most fortunate during this period to have E.S. Pearson as my professor and H.O. Hartley as my advisor. My Ph.D. thesis was titled "A General Distribution Theory for a Class of Likelihood Criteria."

CHAPTER THREE

ICI and the Statistical Methods Panel

WHILE I was a student at the university, my three-month summers were spent working for Imperial Chemical Industries (ICI) where I was later employed. The company was huge and was divided up into divisions located at different places in Great Britain. The divisions included dyestuffs, paints, textiles, pharmaceuticals, heavy chemicals, explosives, and so on.

The first year I was an assistant to Mr. L. R. Connor at ICI's London headquarters, where as a vacation student, I earned four pounds a week. Mr. Connor was a lawyer and was a very precise gentleman. After I had been with him for about a month, he suddenly asked me whether ICI had been paying my four pounds a week. I said that no, I hadn't been paid yet, but I wasn't depending on it to survive. Mr. Connor, said, rather gravely, "That's all very well, but suppose you were to *sue* ICI?" Somehow I didn't think I'd get very far suing one of the largest companies in post-war Britain.

At the time of my first summer, the Statistical Methods Panel, which ICI had established to coordinate the statistical work of its various divisions, had just finished writing the book, *Statistical Methods in Research and Production*, often called "Little Davies" after it's editor, O. L. Davies. It had been written by scientists at ICI for internal use. Someone thought that it would be a good idea to ask Lord McGowan, the company's CEO, to write a few words for the preface. Lord McGowan surprised everyone when he said that in the aftermath of WWII, it was our duty to help

An Accidental Statistician: The Life and Memories of George E.P. Box, First Edition. George E.P. Box.
© 2013 John Wiley & Sons, Inc. Published 2013 by John Wiley & Sons, Inc.

not just ICI but the whole of British industry. So he decided that the book should be published externally and made available to the general public. There was a bit of a flap on, however, because the authors of the book couldn't very well say, "It was good enough for ICI but not good enough for the general public." I was asked to read through the draft and comment. My suggestions were well received, and because of this I was made a joint author of the book, and later a member of the Statistical Methods Panel.

My second year as a vacation student was spent at the Dyestuffs Division in Blackley, near Manchester. The people there offered to make me a salaried employee during my third year as a student, with the understanding that I would join them afterward. The salary they offered was considerably more than the government grant that I had been receiving, so I agreed.

Soon after I joined the Statistical Methods Panel, we had a meeting in London. On the first morning, Harold Kenney, the chairman of the panel, told me that very unexpectedly, our Managing Director would be attending the meeting. There was nothing on our agenda that was likely to be of interest to this VIP, so at about 8:15 a.m., with the meeting due to start at 9:00, Harold came to me, the most junior member of the panel, and asked whether I would make a presentation. Harold knew that I was working on what came to be called, "Response Surface Methods," a way to run experiments to determine conditions that maximized the yield of a chemical process. I was put in a separate room to think, in the few minutes available, how to explain this clearly to this very intelligent but very nontechnical person. I hastily made some notes and hoped for the best. To my surprise, our visitor became quite interested in what I said and asked lots of very sensible questions. So it was a success, and Harold never forgot it. After this I could do no wrong. The other members of the panel were all very senior to me, but during meetings, Harold would always interject, "I'd like to know what *George* thinks about this."

The Statistical Methods Panel met several times a year, and because the various parts of the company were spread all over the country, we needed to find a place conducive to creativity that was readily accessible to all of us by train. We found such a place at Keswick, a beautiful location in the Lake District of northern England, and we met there for periods of three to six days.

Some of the happiest years of my life were the eight that I worked at ICI. Among the things ICI made were synthetic dyestuffs, textiles, and waterproofing and mothproofing agents. Expert teams of highly skilled chemists and engineers developed and improved the complicated processes associated with these products. I quickly got myself involved with them and was able to increase the efficiency of their experiments, both on the full scale and in the lab. Typically a 1% increase in yield could produce huge profits. To help them design effective experiments, I had to know details of the processes and testing methods, so I found myself climbing up and down ladders, talking and arguing every day with technical staff and process workers, and teaching them a little about statistical design and analysis. The woman who brought tea to our desks each morning and afternoon soon tired of taking away my untouched cup, and complained to my secretary, "'E's never 'ere."

In 1955, a young man named Norman Draper worked with me at ICI as a summer student. He recalls riding his motorbike to the job, which paid the paltry sum of five pounds a week. He expected that the job would entail endless data entry. Instead I sent him all over to talk to scientists, get answers to questions, and discuss problems. Norm followed that summer by enrolling in the Ph.D. program at North Carolina, where he studied with Chandra Bose. In 1960, after he graduated, he came to Madison to work at the Math Research Center, and later the Statistics Department.

I turned out a great deal of work for the company, and as a result, my boss said if I wanted to go to meetings or to lectures or anything like that, I should just go. I need not ask permission from anyone. So I went to the afternoon lectures of Professor M. S. (Maurice) Bartlett who was then at Manchester University. Professor Bartlett taught a number of courses, including one on a branch of mathematics called the theory of games and ethical decisions. A number of student athletes enrolled in this class with the mistaken idea that it would increase their prowess in their particular sport. I took his course on multivariate analysis, and his lectures were extremely clear and inspirational, particularly since he used n-dimensional geometry to illustrate the mathematics. The discussions that took place during afternoon tea were most informative as well.

Using these ideas, George Tiao and I subsequently wrote a paper that included the key idea of finding linear combinations of nonstationary time series that are stationary, that is, the idea later called cointegration.[1]

The work that I did at ICI was to help chemists and chemical engineers design and analyze experiments to improve chemical processes; some of the work was done on the laboratory scale, some in the pilot plant, and some on a full scale. Experimentation on the process itself was expensive and difficult. Lab experiments were easy, but the drawback was that you always needed to make an educated guess as to how these small-scale results might apply on the full scale, and we knew that sometimes our guesses could be wide of the mark. As a supplement to this work, therefore, in 1954 I devised a technique called "evolutionary operation," which I communicated to the Board of Directors in a short memo.

The objection to experimentation on the full scale is that combinations of variables would have to be tried that might disrupt the normal operation of the process and produce unsalable products. Evolutionary operation was run on the full scale but did not have this disadvantage because it used Darwin's concept of evolution and natural selection. The changes from the best known process conditions were very small, but they were repeated many times over. The idea is illustrated here for the simple case of just two reaction conditions of temperature and concentration and one response variable percentage yield.

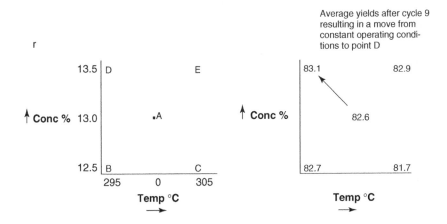

---

[1]G.E.P. Box and G.C. Tiao, "A Canonical Analysis of Multiple Time Series," *Biometrika*, Vol. 64, 1977, pp. 355–365.

It is supposed in this example that the previously best known operating conditions for temperature and concentration were $300°C$ and 13%, respectively, labeled on point A in the left-hand diagram. In the evolutionary operation mode, the process was run for a suitable period at each of the different conditions labelled A–E. The conditions B–E are chosen to be only slightly different from A—not enough to cause problems. But the cycle was repeated a number of times and the results averaged. As soon as there was evidence that some variant was significantly better, this became the origin for a new cycle. In the illustration in the right-hand diagram, after nine cycles of the procedure, the data are starting to show that a slightly lower temperature with a slightly higher concentration would give an increased yield. The ICI board was reluctant to have this idea published but eventually allowed it, in 1957.[2]

In 1998, the Institute of Electrical and Electronics Engineers (IEEE) produced an anthology of the papers and reports leading up to what became a method for computer calculation called *evolutionary computation*. One of the earliest of these was the one on evolutionary operation that I published. For this the IEEE awarded me the Evolutionary Computation Pioneer Award in 2000.

At ICI, in addition to helping to improve chemical processes, a massive effort was needed to test the products thoroughly: dyestuffs, detergents' waterproofing agents, artificial leathers, and many others. How close to standard colors were our dyestuffs? How waterproof were fabrics treated with our products? How resistant to wear was our artificial leather? To obtain the answers to such questions, a sample of the material to be tested was compared with a standard that had to be matched or, in appropriate cases, exceeded. There were ingenious instruments and machines for carrying out the tests. Once more these provided a golden opportunity to employ Fisher's experimental designs. For example, in the Martindale wear-testing machine, four small pieces of fabric, one of which was the standard, were placed in holders and rubbed at a fixed pressure against abrasive emery paper. The loss in weight after a 1000 cycles of the machine was a measure of resistance to wear. Thus, three sample products could be compared with the fourth, which was

[2]G.E.P. Box, "Evolutionary Operation: A Method of Increasing Industrial Productivity," *Applied Statistics*, 1957.

the standard. There were obvious difficulties. For example, four holders were used. Were there differences due to holders? The holders were in four different positions on the machine. Were there differences due to positions? In each run of the machine, the emery paper abrasive needed to be changed. Were there differences due to different emery papers? How could we allow for such differences? The subsequent figure shows part of the replicated Hypergraeco Latin Square design with the numbers showing wear after four cycles of 1000 revolutions. This kind of design is due to Fisher, an elaboration of the Latin Square design mentioned earlier.

		Positions				Replicate I
		P_1	P_2	P_3	P_4	
	C_1	$\alpha A1$ 320	$\beta B2$ 297	$\gamma C3$ 299	$\delta D4$ 313	Cycles: C_1, C_2, C_3, C_4
	C_2	$\beta C4$ 266	$\alpha D3$ 227	$\delta A2$ 260	$\gamma B1$ 240	Treatment: A, B, C, D
Cycle	C_3	$\gamma D2$ 221	$\delta C1$ 240	$\alpha B4$ 267	$\beta A3$ 252	Holders: 1, 2, 3, 4
	C_4	$\delta B3$ 301	$\gamma A4$ 238	$\beta D1$ 243	$\alpha C2$ 290	Emory paper sheets: $\alpha, \beta, \gamma, \delta$

Hyper-Graeco–Latin Square—Wear Testing[3]

Using the design, it is possible to eliminate the effects of position, cycle, holder, and emery paper and thus to obtain very accurate comparisons with the standard. I very much enjoyed solving complex puzzles associated with such experimental designs.

One thing I greatly missed after I left ICI were the stories and jokes that went around. The standard was high, and it seemed a new joke surfaced almost weekly. Harold Kenney, for example, was a source of good stories. Some of these were about a friend of his called Hetteridge, who had served in the trenches during the First World War and was one of the few people who seemed to have enjoyed it. He was quite blood

---

[3]G. E.P. Box, J.S. Hunter, and W.G. Hunter, *Statistics for Experimenters*, 2nd ed., Wiley, New York, 2005, p. 163.

thirsty. He had tried very hard to rejoin the Army at the beginning of the Second World War, but he was too old.

After the German blitzkrieg that had defeated the French in a matter of weeks, the authorities in Britain were concerned about the possibility of an invasion that might perhaps use parachute troops. So the Home Guard [sometimes called "Dad's Army"] was formed. Able-bodied men who were too old to serve in the Army, Navy, and Air Force joined the Home Guard. At first they didn't have very much in the way of weapons, so they had drills with shovels and rakes temporarily substituted for guns. Hetteridge was a major in the Home Guard. He was extremely keen and became well known for the efficiency of his Home Guard unit.

There happened to be a regiment of the regular Grenadier Guards stationed nearby, and their colonel asked Hetteridge over for dinner in the officers' mess. Hetteridge arrived in his First World War uniform with two Lugars (German guns) stashed in his belt. The colonel met him at the door but pointed out that one did not take weapons into the officers' mess. So reluctantly Hetteridge hung up his belt with his guns in the hallway. The dining room had French windows overlooking a pristine lawn and flower gardens. The conversation at lunch was about the fall of France and the danger from parachutists. A lieutenant was speaking about this then novel form of warfare, and looking out onto the lawn, he said, "There could be a parachutist coming down on the lawn right there and what could we do about it?" Hetteredge felt inside his coat, produced another Lugar, and said, "I'd shoot the bastard."

Another Hetteridge story concerned the fact that among other things, he was a lay preacher. Unfortunately, one of Heterridge's characteristics inherited from WWI was his use of "bad language." This kind of vocabulary does not translate easily from English to American and vice versa. For example, while the word "bugger" is fairly innocuous in American, in English, it is not. In particular the expression "bugger off," meaning "go away!", is not used in polite society. Now immediately after the fall of France, a large proportion of the British Army was rescued at Dunkirk and successfully brought back to England. The following Sunday, Hetteridge based his sermon from the parable of the seven lepers in which Jesus cured seven lepers of their leprosy, but only one returned to thank him. With great emotion, Hetteridge said, "We have just witnessed a *miracle*! The great majority of our army has been rescued in the face of the enemy. Yes, a *miracle*, and what are we doing? Are we on our knees thanking God? No! Things are just happening as usual! It's just like the story of the seven lepers—*seven* were cured of their leprosy and only one returned to thank Him. *What did the other six do?* Why! They just buggered off!"

Jessie and I lived in Sale, about 12 miles from Blackley where I worked. A number of people from ICI lived nearby, so we jointly hired a deluxe coach to take us to and from work. At times these trips offered unexpected entertainment. After the war, new cars were almost impossible to get, so it was regrettable when one morning a brand new car was badly damaged in a collision with our bus. While we waited for the police to come, the driver of the new car sat down disconsolately on the sidewalk. A small crowd of somewhat grubby children who, I suppose, were on their way to school, quickly gathered. After a time someone in the bus must have been telling a joke because a small urchin peering through the door of the bus and greatly impressed by the damage that our vehicle had done to this beautiful car, said to the hapless car owner, "Here, mister, *this* one's laughing!"

On another bus journey, I was sitting next to a very dapper little man who said to me, "I understand you're going to America." I said, "Yes." "What are you going to do with your house?" he asked. I replied that we intended to let it. He said, "Our family did that when we went abroad and we had a most *unfortunate* experience." "How so?" I asked. "Well," he replied, "the first people who rented were not so bad, although they did steal the silverware. But the people who rented after them were worse. It seems they were running a disorderly house." He then went on to say, "Our house rather lent itself to that because it had so many bedrooms."

Later I was in a car pool. In those days, everyone heated their houses with coal. This and the climate inversions that occurred in Manchester periodically produced "pea soup" fogs. These fogs were like nothing I have encountered anywhere else, and when they happened, we would be told to quit work early and try to get home. The visibility on such occasions was literally no more than four feet. To navigate we had someone, who was just visible to the driver, walk by the side of the car and someone else, who was just visible to the first man, walk on the sidewalk.

I remember on one such expedition the road must have turned sharply. But we didn't, and together with several other cars, ended up in someone's garden. The owner of the house was jumping around solidly cursing us as we maneuvered across his flower beds, backing and turning in order to find our way out.

At some point, "smokeless fuel" became available, the burning of coal was banned, and the pea soup fogs disappeared.

"When I use a word . . . it means just what I choose it to mean—neither more nor less."

CHAPTER FOUR

George Barnard

GEORGE BARNARD was a brilliant mathematician (Figure 4.1). He had come from a poor family and on his own merits went to Cambridge to the best math department in Britain. For some reason, when you were with George Barnard, interesting and unexpected things tended to happen.

Imperial College, where George had his office, was one of the institutions built by Prince Albert with the proceeds from the Great Exhibition of 1851. At one time, its rooms must have been very large, but expansion of the faculty had, over the ages, produced many extra walls. Consequently, it now had a number of very narrow rooms with very high ceilings. One of these was George Barnard's office. I remember one particular day when I had gone to see George to discuss a problem, but we never really got to it. Soon after I arrived, George said, "You must find it stuffy in here. I'll open a window." That window must have been closed for a very long time for it proved impossible to move it. George's secretary, Miss Mills, who was used to coping with all kinds of emergencies, brought a screw driver and a hammer and every possibility was tried. Eventually, after a lengthy struggle, the window was opened. But then, almost immediately, two pigeons flew in.

There was a passageway that passed through George's office, and the pigeons settled on its roof. For a long time George, his secretary, and I stood on chairs with pointers and other instruments trying to get the pigeons to go back out of the window. In the end we were successful, but

An Accidental Statistician: The Life and Memories of George E.P. Box, First Edition. George E.P. Box.
© 2013 John Wiley & Sons, Inc. Published 2013 by John Wiley & Sons, Inc.

FIGURE 4.1
George Barnard making tea.

then George said, "Let's go to lunch." We strolled down the road, no doubt very deep in conversation, and we came to a pedestrian crossing. Now in England, the strictly enforced rule was that no vehicle was allowed on a crossing if a pedestrian was already anywhere on it. But on that particular day, there was a very expensive car, all chromium plate and headlights, impatiently edging its way onto our crossing. The driver looked very angry, for he, no doubt, regarded our progress on the crossing as too slow. When we came level with the car, George became aware of this vehicle inching its way onto the crossing and he immediately stepped in front of the car, banged with both hands on the hood and shouted, "*Back! Back!* Get off the crossing. *I'm not going to move until you do!*" Eventually of course the motorist did retreat, but he didn't much like it.

Many of my most memorable experiences with George Barnard occurred when we were both members of the Research Committee of the Royal Statistical Society. The committee was responsible for spotting talent and finding speakers with something new to say in "read" papers for the Society. George had become a consultant to ICI, and from time to time, he visited the Dyestuffs Division and was familiar with what we did there.

I had been working in particular with a chemist, Dr. K. B. Wilson, on a general technique called "response surface methodology" for improving a process. George suggested that Wilson and I write up our work so that it could be presented and discussed as a read paper at a Research Committee meeting. This idea astonished me at first, but eventually we wrote the paper and submitted it. One of the referees, a very well-known statistician who had been a student of Fisher,[1] said the paper should be rejected. He felt so strongly about this that he threatened to resign from the Society if it was accepted. However, in the end, in some way that I have never discovered, his objections were overcome and the paper was presented and discussed and duly appeared in the *Journal of the Royal Statistical Society*.[2] The paper was a great success and, as I will later explain, was the cause of my first coming to the United States.

When I was elected a member of the Research Committee, there were about eight of us and we came from all over England, Scotland, and Wales. George drove a van called a "Dormobile," which could carry a number of people. Typically, after the meeting in London, he took all of us to our appropriate railway stations from which we could travel home. At that time, a topic about which there was a great deal of dispute (sometimes heated) concerned theories of statistical inference. Among these were the Neyman–Pearson theory, Bayes' theory, likelihood, and fiducial inference.

This last was a gallant (but I think unsuccessful) attempt by Fisher to avoid certain difficulties encountered with the others. Now George was a great admirer of Fisher and thought he could do no wrong (indeed it did turn out that Fisher was right about most things). However, Dennis Lindley, who was one of George's passengers, believed that *Bayes* was right, and Fisher was wrong, and expressed his opinion forcibly. George was incensed and went through two red lights. When the cries of

---

[1] Fisher himself liked our ideas and suggested the name "Technometrics" for a new journal that might publish this kind of work.
[2] G. E. P. Box and K. B. Wilson, "On the Experimental Attainment of Optimum Conditions," *Journal of the Royal Statistical Society*, Series B, Vol. 13, 1951, pp. 1–45.

"Oh my God!" had died down, we all took a solemn vow never again to discuss theories of inference while traveling in George's dormobile.

David Cox (later Sir David Cox) was a member of the Research Committee at the same time that I was. We often heard, "Ha Ha! Box and Cox! You must write a paper together." So we thought we would do that, but what should the paper be about?

In the late 1880s, there had been a comic play called "Box and Cox." Box had a night job and Cox had a day job. The woman who kept the house where they boarded got paid double by renting a room to Box during the day and the same room to Cox at night. We got rather fed up with the teasing, but knowing the story, we decided that obviously our paper should be about transformations.[3] In this spirit, we deliberately included two different derivations of our results, one using likelihood and the other Bayes. At the meeting, the discussants tried hard to find out who wrote what, but we were not about to tell them. From a practical point of view, it didn't make much difference anyway.

After our research meetings, we usually took the speaker out to dinner at an Italian restaurant called Bartorelli's. At one particular meeting, George had strongly criticized the speaker on a specific point. As we were walking over to the restaurant, the speaker said to George, "I thought you were rather hard on me tonight. I heard you say the same thing yourself not more than 12 months ago." George looked surprised and slightly indignant and said, "You don't imagine I'd necessarily think *this* year what I thought *last* year, do you?"

In 1953, when I first came to the United States, Senator McCarthy was very much in the news, so I made a point of going to Washington to observe the activities of the House Un-American Activities Committee. I witnessed first hand how the epithet "communist" could be used to destroy a person's reputation. During this time, it happened that many statisticians on the American side of the Atlantic wanted to get a chance to talk with George Barnard. But there was a problem. Immediately before WWII, George had spent some time at the math department at Princeton. The FBI claimed that while he was there, he had formed a Communist cell.

---

[3] G.E.P. Box and D.R. Cox, "An Analysis of Transformations," *Journal of the Royal Statistical Society*, Pt. B, Vol. 26, 1964, pp. 211–252.

As a young student, George, like many, had sympathies with the Communist Party. Later, when I knew him, he was liberal in his outlook and not exceptionally so. What the "cell" at Princeton had been is unclear, but the FBI refused him a visa numerous times. There was an outcry by many U.S. statisticians who worked very hard to show that George had statistical ideas they needed to know more about, and that he did not constitute a danger to the United States. Over the next year or two, there was much writing to senators and various attempts to get him a visa.

At one point, we got around the difficulties by having George present a lecture course on the Canadian side of Niagara Falls, and many American statisticians came to hear him. But George had greatly enjoyed being in the United States and had looked across the falls with nostalgia. Eventually the FBI said they would give him a visa if he would give them the names of all who had been in the group at Princeton. This, of course, George refused to do, but in 1961, when George was invited to visit the National Institutes of Health, the FBI agreed to allow him in with a "waiver." I think someone presented a document to George, covered it with his hand, and said, "Don't try to read this, just sign it." And eventually George did.

George's problems with the FBI did not end, however. From 1975 to 1981, when George spent a part of each year as professor of statistics at the University of Waterloo in Canada, difficulties entering the United States still plagued him.

The adventures with George went far beyond statistics. One of the most memorable houseboat trips I organized was with the Barnard family. George and Mary had five children, so we needed rather a large boat to provide sleeping accommodations for everyone. We usually started the trip near Windsor, traveled the Thames to Oxford, and then back again, going about 160 miles in all. Every few miles along the river there was a lock. The lock keepers were ex-Navy men, and we were expected to observe proper protocol. If you were going upstream, you would wait for the lock gates to open, then you moored the boat with the appropriate length of rope, and without bumping the sides of the lock, kept the boat steady as the lock filled up. Most lock keepers were friendly and helpful and would sell you fresh produce from their gardens. No one was in a hurry, and we spent about three days going up to Oxford and another three days coming back.

The children were very watchful of their prerogatives. If, for example, one child was permitted to steer for a bit, the other four children watched very carefully, and noted the time the amateur helmsman was at the wheel. Then each insisted that they should be allowed to do the same and for *exactly* the same amount of time. If their time was half a minute short, he or she protested that they had "not had a *proper go.*"

On a different occasion, we went very early, in April. One chilly morning, I started the engine just as George released the mooring rope, which wound itself very tightly around the propeller shaft. The Thames is cold in April, but there was nothing to do but to take turns diving under the boat with a penknife, and bit by bit we cut through the layers of rope until our breath gave out. Neither of us had brought swimming gear so Mary, George's wife, leant each of us a pair of her drawers. Although it took a long time, we were finally successful, but both of us were frozen to the bone. Mary, who was a Scot and a nurse, gave each of us a half-tumbler of whisky, which made us warmer but a bit unsteady.

There were many places to tie up along the river and to explore the surrounding countryside or buy provisions. On one occasion, we got off the boat, walked across some fields, and came to a village with a very small pub. The person managing the pub seemed to think that we were posh people, so he opened up another room for us that was reserved for special occasions. There was a piano there, and I sat down and played. Soon after our arrival, a large bus of people with strong cockney accents arrived and it was clear that they were out celebrating. One of them peeked into the room and asked, "Can *anyone* come in here?" When we told them yes, apparently so, I was soon joined around the piano by a boisterous group, and we sang songs like, "Knees Up Mother Brown," a tune that inspired others in a vigorous dance around the floor.

On one occasion when I was making a brief visit to England from the United States, I was anxious to talk to George, but upon inquiry, it turned out that he was chairing a meeting of the Industrial Applications Section of the Royal Statistical Society at Jesus College, Oxford. I drove to Oxford, and we had a chance to talk. I arranged to stay at the College, but I kept in mind that I had to get up very early the next morning to drive to Heathrow to catch a plane. The person who had organized the meeting was Miss Joan Kean, the secretary of the Industrial Applications

Section who was staying at the College. I told her of my need for an early start in the morning, and she lent me her alarm clock.

The following morning I was up and dressed at 5 a.m., but I found I could not get out of the college. There was an extremely large and formidable locked door and a high wall that had been there since the year 1500 or so. So I started to look for another exit. I went all around the college twice but could not find a way out. There were stairs leading down to some sort of a dungeon, so I wandered about down there, but again there was no exit. It finally occurred to me that over the centuries the college had wanted to keep the undergraduates inside and they had done a first-rate job.

I then remembered that I had to return the alarm clock to Miss Kean, and that this would be a good opportunity to consult with her. I knocked on her door, and she eventually appeared in pajamas and dressing gown. I told her my problem. She said, "There surely must be a way out. I'll come with you." So we repeated the search I had already made without success.

After some time, we eventually found a ladder and discovered that there was an outer building attached to the wall, so we climbed up onto the roof of this building and we were just lowering the ladder down to the road on the other side when we saw a policeman looking up at us. We must have looked rather strange: I was carrying a suitcase, and she was wearing pajamas and a dressing gown. I said to the policeman, "I am Professor Box of the University of Wisconsin, and this is Miss Joan Kean, the secretary of the Industrial Applications Sector of the Royal Statistical Society who is helping me. I need to leave and catch a plane, but there seems to be no way out of this college." I think this kind of thing must have happened before, for the policeman didn't look very surprised. After a little thought, he said, "Right 'o. I'll 'old the ladder and 'elp you get down." So I said goodbye to my companion on top of the wall and was soon on my way to Heathrow.

Later when I was in England, I told this story at a dinner party. As we were putting on our coats getting ready to leave, a very, very proper English lady took me to one side and said, "George," and I said, "Yes, Elizabeth" and she whispered to me, "I just wanted to let you know that I think you were *quite right* to introduce Miss Kean to the policeman."

I started smoking when I was first in the Army. When it came time for a break, the sergeant would say, "Smoko." There were few, if any, abstainers. Later, when I met George Barnard, he encouraged me to give up the habit because he had just become aware of the research done by Bradford Hill and Richard Doll that revealed the possible link between smoking and lung cancer. Their 1954 paper, which has been unjustifiably overshadowed by later studies in the United States, gave conclusive proof of the link.[4]

Hill was a brilliant medical statistician. In the 1940s, he conducted the first randomized clinical trial in a study of the effect of streptomycin in the treatment of tuberculosis. Fisher, who had pioneered randomization in field trials at Rothamsted 20 years before, was on friendly terms with Hill. In 1954, he proposed that Hill be made a member of the Royal Society, and Hill became a fellow that year. But Fisher was also an avoid smoker but by the late 1950s, Fisher openly criticized the smoking study. He argued that correlation, which *had* been proved, and causation, which had *not*, were not the same thing—for example, a genetic factor in an individual that predisposed him to smoke might also predispose him to cancer. More generally, x may be correlated with y because *both* are correlated with z. The relationship between Fisher and Hill cooled considerably, but the causative evidence for a link between smoking and cancer grew stronger as Hill and Doll's study continued. I found the evidence compelling, and with George Barnard's encouragement, I stopped smoking.

Hill had a great sense of humor. When he gave his first address as president of the Royal Statistical Society in 1950, he explained that he hadn't known what to say, so he had consulted the addresses of past presidents to get some ideas. He saw that a very early past president (the RSS was 150 years old) had received a request that the Society determine the amount of horse manure deposited each day on the London streets. The president had written a careful letter in which he made it clear that while members of the Society would be glad to *analyze* the data, they should not be expected to *collect* it.

---

[4]The paper was issued as part of the British Doctors Study conducted from 1951 to 2001. See R. Doll and A.B. Hill, "The Mortality of Doctors in Relation to Their Smoking Habits: A Preliminary Report," *British Medical Journal*, Vol. II, 1954, pp. 1451–1455.

When I first knew George Barnard, he had a house on Barnes Common by Thames. Especially at night, the house had a miasma of mystery of the late Victorian period. The bathrooms in particular dated from that time, and I was especially fascinated by the shower in one of them. This was a glass box, and when you closed the door and turned it on, you were attacked by multiple squirters that came at you from all directions. These had the peculiarity that some were boiling hot while others were stone cold. It was an excruciating experience. I think I must have learned how to turn the shower on but not how to turn it off. Certainly on the one occasion when I used it, I must have left it running all night. The Barnards were very nice about it, but everywhere downstairs was flooded with water. Armed with mops and buckets, it took all the adults and children most of the day to get it reasonably dry. I think George forgave me, but I suspect that Mary never quite did.

Later, George and Mary Barnard had a beautiful house known as "Mill House," close to Brightlingsea, on the Thames estuary. They spent much time and labor fixing it up. I remember visiting unexpectedly when it had been raining hard, and the fixing up process was in its early stages. In particular, the roof of the passage that led to the kitchen leaked. George was away in Canada, and Mary was thoroughly cursing him.

Brightlingsea had been a favorite haunt of King Edward the Seventh, who kept his yacht there, and it was still a favorite place for sailors. On one occasion, I was visiting with my two children who were quite young, Helen being perhaps six, and Harry four. At some point, George had acquired a small boat with an outboard motor. The children and I climbed into the boat with George who was having trouble starting the motor. He gave a very hard tug on the line, but because he was concentrating on the engine at the back of the boat, he didn't see where he was going, and suddenly the motor roared into life. We charged a bollard at great speed, and the children were thrown into the bottom of the boat. After a bit, George tried again. But this time, we hit someone else's boat with the same result. Picking herself up once more, Helen remarked to Harry, "I hope it doesn't go on like this *all day.*"

I stayed with George and Mary at Mill House from time to time, enjoying the lovely garden and fish pond. The house contained various gadgets due to George. I remember in particular an elaborate system of rubber tubes in our room that came with typewritten instructions as to how, by a series of maneuvers, it was possible to get hot water.

One day I went for a walk with George along the bank of the Thames and George suddenly said, "Do you realize that we are at a National Monument?" I could see nothing but a great deal of mud, but George explained that this was the site of a Roman Oyster Bed. I had no reason to doubt him.

"The time has come, 'the walrus said,' to talk of many things.
Of shoes and ships and sealing wax, of cabbages and kings..."

CHAPTER FIVE

An Invitation to the United States

I was happy with what I was doing at ICI, and I had had no thought of academia. But in the course of solving practical problems, I had come up with a number of ideas for the development of statistical methods and had published them. In 1952, I was surprised to receive a letter from North Carolina State University at Raleigh, which had established one of the first departments of statistics in the United States.[1] The letter was from Miss Gertrude Cox, who famously ran the Institute of Statistics with departments on both the Raleigh and Chapel Hill campuses. It was an invitation to spend a year at Raleigh as a "Visiting Research Professor."

I later found out how this came about. J. Stuart Hunter, then a graduate student at Raleigh, had worked during the summer vacation at the Army Research Office (ARO) (Figure 5.1). Stu had seen the paper I had written at ICI with K. B. Wilson in 1951, which concerned the experimental determination of optimal process conditions. He showed this to Frank Grubbs, who was in charge of ARO, and Frank proposed to Gertrude that she use some ARO funds to invite me over. The ICI board of directors gave me a year's leave of absence, but they made it clear that they wanted me back.

---

[1] Then called North Carolina State College, part of the University of North Carolina system, which included the campus in Chapel Hill.

An Accidental Statistician: The Life and Memories of George E.P. Box, First Edition. George E.P. Box.
© 2013 John Wiley & Sons, Inc. Published 2013 by John Wiley & Sons, Inc.

Investigating a design using the fewest possible experiments in such a position as to yield the maximum information about an unknown surface, such as crop yield as a function of various fertilizer combinations, or a yield in a chemical reaction involving many reagents. (L. to R.) G. E. P. Box, R. J. Hader, and J. S. Hunter.

FIGURE 5.1
George Box, Ralph Hader, and Stu Hunter.

At that time, I had not submitted my Ph.D. thesis. People told me that, although this didn't matter much in England, in the United States, everyone was supposed to have a Ph.D. I had already written my thesis, much of which had appeared in already published papers. So I had my examination just a few days before I was to leave for the United States on the *Queen Mary*. My examiners, E. S. Pearson, H. O. Hartley, and M. S. Bartlett didn't mention statistics but chatted with me about the comparative advantages of going to America by air or by sea. When I asked if I'd passed, they said, "Yes, of course."

At ICI I was attached to a group called the Miscellaneous Chemicals Division. It included various activities such as X-ray Crystallography that didn't fit anywhere else. Our boss was Dr. S. H. Oakeshott. When the board decided I could go to the United States, Oakeshott asked me to come and see him. He said, "Now we'll need to give you some money to get to North Carolina because it's on the west coast of the United States and it's a long train ride." I told him, "No, it's in the east." We had a discussion, and finally I remembered the words to the song "The Chattanooga Choo Choo." I sang some of the words to him:

> *You reach the Pennsylvania station 'bout a quarter to four,*
> *You read a magazine and you're in Baltimore*
> *Dinner in a diner*
> *Nothing could be finer*
> *Then to eat your ham and eggs in Carolina*

[CITATION: Harry Warren and Mack Gordon, 1941; first recorded by Glenn Miller and His Orchestra in 1941.]

I said, "So you see, it's got to be on the east coast." He was a very nice man of the "old school tie" type, and he didn't seem to follow me, but eventually we got out a map and he realized that I was right.

Jessie and I left Southampton on the *Queen Mary* in the middle of winter. When we first arrived at the dock, there was great excitement on board with cameras and lights everywhere. It turned out that Sir Winston Churchill was crossing to see President Eisenhower. Two or three members of his family and a number of important people were there to see him off. There was much flashing of cameras and general commotion, but eventually he was ushered away. Jessie and I had been watching all of this from close to the top of the gangway. We too decided to move on, but didn't know our way around the ship. We found ourselves walking down a narrow corridor where we could just see two people walking toward us from the other end. It turned out to be the Captain and Churchill. To get past them, we had to go stomach to stomach.

Not long after we put to sea, we noticed members of the crew setting up ropes by all the stairs and passages. I asked if they were expecting rough weather and they allowed that this might be the case. The steward suggested that the best chance of avoiding sea sickness was to stay up on deck in the open air. On deck where there was a bar and, usually, dancing. After it got rough, I saw one couple attempting to dance, but the floor was rising and falling at strange angles and they had to give it up. One interesting phenomenon was to see a drink creep up one side of the glass and then down and up the other side.

The storm became dramatic. I got up on deck and wedged myself in the highest place I could find, facing forward. We cut through huge waves, the enormous ship diving deep into the water and then flying up with an immense volume of water breaking over its prow.

The remainder of our crossing was rough, but we finally got to New York, where the authorities came on board to check passports and visas and inspect the chest X-rays that everyone had to present before they were allowed off the boat. We disembarked and had breakfast in New York. This cost eighty cents, an amount that at the time seemed almost ruinous when converted into British currency.

On the plane flying down to Raleigh from New York, I got into conversation with my neighbor, an Egyptian student. By coincidence we were staying at the same hotel, and we met the next day for breakfast, where we discussed the many different kinds of English spoken in the world. As if to illustrate the point, when I asked the waitress for ham and eggs, she said, "Huh?" I repeated my words with the same result. Finally I *pointed* to the items I wanted on the menu. My Egyptian friend watched this performance with wide-eyed surprise and then remarked, "If *you* cannot understand them, how can I?" When the waitress brought my breakfast, in the corner of my plate was a mess of some kind. I took it back to the counter and said, "I didn't order this." She exclaimed, "Why them's yer grits!" Like french fries in Australia, you got them whether you wanted them or not.

I was not the only Englishman who had initial difficulty with certain American traditions. When I got to Raleigh, I heard a story about a very large park that was nearby, and during the war, it was made available to British submarine crews to rest up. A number of southern ladies had volunteered to help, and among other things that were offered to the soldiers was iced tea. This was a beverage that was totally unknown in England. It took some time to discover that out of sight of the ladies, the crew had built a fire and used it to heat up the tea, thereby making it drinkable.

After settling into our new quarters in North Carolina, I became acquainted with Miss Cox (Figure 5.2). She was surprising: What you thought you saw was a pleasant, middle-aged lady who liked to tend her garden and bake cookies. Despite of appearances, she was a ball of fire. She had been a researcher and assistant to Professor George Snedecor at Iowa State, where the latter had founded the first department

FIGURE 5.2
Gertrude Cox.

of statistics at a U.S. university. Snedecor had been asked to suggest a person suitable to head the Department of Experimental Statistics that was being established at North Carolina State, and he came up with the names of five men. When Miss Cox asked, "Why don't you recommend me?" he did, and in 1940, she went to Raleigh. When I got there in 1953, she had organized and was running the Institute of Statistics. This included a department at Chapel Hill, concentrating on theory, as well as the Raleigh department, concentrating on more practical things. She had the support of the governor and was also involved in founding the "Research Triangle," a joint effort from Raleigh, Chapel Hill, and Duke University. She was one more example demonstrating that a woman can be an inspired leader and a super competent administrator.

The following story illustrates her remarkable talents. She had an offer of half a million dollars if she could find matching funds for the other half a million. She asked the help of a famous industrialist who said he would have no difficulty in raising the matching funds. There was a time limit of 12 months, but at the end of 9 months, he had failed to get the money. She said, "Okay, will you let me try?" She approached the faculty and the graduate students and explained her problem. She said, "I want each of you who is doing practical research to write up a ten-minute talk. We'll make the rounds of the big shots and see if we can raise the money." I helped by contributing a piece on efficient experimentation and was moved to write a little ditty:

Here we come with whistle and flute,
Collecting for the Institute

She got the money!

What was taught at Raleigh and Chapel Hill was Fisherian statistics: the analysis of variance and other ideas due to Fisher. My research on response surface methods was a natural extension of Fisher's concepts applied, however, to technology rather than to agriculture. My ideas were known to Fisher who was delighted that his kind of statistics was being developed in new areas.

Unfortunately there was a graduate student who went around saying that people were wasting their time studying Fisherian statistics that had (in his mind) been superceded by response surface methods. Stu Hunter found out about this. He told me that the faculty thought that the student's opinions had come from me, and they were angry about it. Each month Gertrude had a session at her house where matters of interest were discussed, and Stu warned me that I was to be "called over the coals" at the next session. The published account of response surfaces was rather mathematical, but the basic ideas were sensible and easy to understand, so very quickly, I wrote an applied paper called "The Exploration and Exploitation of Response Surfaces" specifically for Miss Cox.[2] She studied it before the meeting, and when the criticism started,

---

[2] G.E.P. Box, "The Exploration and Exploitation of Response Surfaces: Some General Considerations and Examples," *Biometrics*, Vol. 10, No. 1, 1954, pp. 16–60.

she intervened and said that she thought these ideas were excellent and didn't in any way conflict with those of Fisher or with what was being taught in the department. She added that she planned to publish my paper in *Biometrics*, the journal of which she was editor.

The story goes that at one point Miss Cox invited both Sir Ronald Fisher and Frank Yates to present a series of lectures at North Carolina. She worked Fisher pretty hard, so he was quite glad when it was Independence Day and he was able to take off on his own with his butterfly net. But one of the graduate students had seen him and thought that now was his chance to get close to Fisher. The student found him and said, "It's a fine day." Fisher said, "Yes." The student said, "We get a day off—it's Independence Day." Fisher said, "Yes." The student then said, "I suppose you don't celebrate that in England." At that point Fisher turned on him and said, testily, "No, perhaps we *should!*"

Howard Hotelling was a distinguished statistician at Chapel Hill, and at times, he could be a bit pompous. One day he came to the park to attend a department picnic driving a large new car. He asked some graduate students to drive it to a safe parking place where it would not be scratched. There were a lot of trees, and they took a long time to park it. Sometime later, the professor needed his car, but the original team of graduate students was no longer there, so another group took on the job. After a considerable time, they confessed themselves beaten. The car was completely surrounded by trees, seemingly dropped from the sky. For a long time, they were completely baffled (and this, remember, was a group of scholars, all candidates for the degree of doctor of philosophy at one of the finest universities in the country). After a considerable time, there was a breakthrough, and there were those present who felt that they should be awarded their doctorates without further delay. But no theses had been produced with pages that met the exacting requirements of proper margins and numerous signatures, so that plan fell through.

It seemed to me then, as it does now, that something new in statistics most often comes about as an offshoot from work on a *scientific* problem. With this in mind, during my year at Raleigh, Stu and I helped a chemical engineer with his investigations. His name was Dr. Frederick Philips Pike. He had a great sense of humor, and we got on famously. He told me that he was a distant relative of Lieutenant Pike who was the original surveyor of Pike's Peak.

Pike said that as a young man he had been anxious to visit Pike's Peak and had managed to get close to the top in his old car. He was starting on the return journey when someone opened the opposite door and sat down in the passenger seat. The newcomer wanted to be taken down the mountain, and it was evident that he was very drunk. So Pike took him along, and as they proceeded, he told Pike about his recent history. He had left home three days previously, and before he had left, he had had a violent quarrel with his wife in which neighbors and relatives had been involved.

They proceeded for some way in silence, and about halfway down the mountain, his passenger asked Pike what he did. Pike said, "I'm a mind reader." The drunk said he didn't believe him, so Pike told him the same story that the drunk had previously related to him. As Pike suspected, the passenger had completely forgotten he had told him about this, and he began to eye Pike with deep suspicion. Finally, when Pike had driven him home, his passenger left the car, but then came back very angry and said, "I know what's been going on. There's only one way you could know all this. You must have been carrying on an affair with my wife." At that point, he seemed dangerous and Pike quickly drove off.

In Raleigh, every morning on the radio, almost everyone listened to a disc jockey called Fred Fletcher. He was great fun. I remember, for example, that he would often play a request that was a "commercial" for Grandma's Lye Soap." One verse I remember was about someone who

Suffered from ulcers, I understand
Swallowed a cake of Grandma's Lye Soap
Now has the cleanest ulcers in the land . . .

[CITATION: Song goes by same title, John Standley and Art Thorson, 1952.]

If you had anything to sell, Fred would advertize it for you. I remember that for some days, there was a donkey tethered outside the station and Fred would ask for and get food for it while waiting for a purchaser. One day he advertized a piano, and I bought it for 40 dollars. It was thick with beer stains, but it didn't sound too bad.

My office was downstairs in Patterson Hall, which was where most of the graduate students were. Some of their lectures were somewhat obscure, so they would come to me for help and we became friends. Halfway through my year in Raleigh, I had to move and I asked Sid Wiener, one of the graduate students, if he could recommend a mover. (Sid always wore a cigar and was from Brooklyn.) He said, "Ya don't want to go to a movah, Professah. We'll move ya, we'll move ya." And he organized a crew of fellow students and rented a truck, and all went well until we got to the piano. Unfortunately my new living quarters were upstairs and the stairs contained a right-hand turn. Try as they would, the students could not get my piano up the stairs and around the corner. Their gallant attempts reminded me of the famous picture of the three marines with the American flag, and they were reluctant to admit, especially to a visiting Englishman, that it couldn't be done. In the end, we took the piano to my office in the basement of Patterson Hall at the university.

I remembered that when my father had sometimes felt a bit down, he'd say, "Let's go and have a tune up!" He particularly liked to play marches. I did too. So one particular day without thinking too much about it, and forgetting I was in the South, I played "Marching through Georgia." A janitor quickly opened my door and said firmly, "We *don't* play that song down here."

It soon became clear that to get around the campus at Raleigh, I needed some kind of conveyance. When I suggested buying a bicycle, the graduate students shook their heads and told me that, like everyone else in the United States, what I needed was a car (Figure 5.3). Among them was a student from Egypt called Alex Kahlil. I told Alex I was planning to buy a second-hand car. He said that he would help me but that he was supposed to take an exam that afternoon. He suggested that if I called his professor and told him I needed Alex's help, perhaps the professor would postpone the exam and Alex could help me buy the car. The exam was postponed, and under Alex's direction, we found a car that we thought might be okay at, let's call him, Dealer A. Alex asked to take it for a trial run; however, on the way back he took it to Dealer B, who also wanted to sell us a used car. But Alex said, "We've got this car to try from Dealer A. Would you check it out for us?" Dealer B did so and told us a number of things that might be wrong with it. Then we went back to Dealer A, and we told him about these possible defects and suggested that he might lower the price, and so on and so on. I think three dealers were eventually involved in this way. After an afternoon of haggling, we got a pretty good car at a bargain price.

The graduate students at Raleigh took some of their courses at Chapel Hill, and they took it in turns driving there in one of the university cars. A story that went the rounds was about the time when Alex drove the car. His driving was unusual in that under his guidance, the car tended to oscillate somewhat from side to side. After a bit, a police car started to follow; I think the policeman might have suspected, incorrectly, that the driver was drunk. He told Alex to stop and asked for his driver's license, and sniffed around a bit, but everything seemed to be in order. He said to Alex, "The reason I stopped you was that you were veering from side to side." Alex said, "I'm sorry, officer, but as you can see, I'm not a native of this country. I'm Egyptian." The police officer said, "Yes." Alex said, "Well in Egypt, of course, you learn to drive a camel, and sitting on the back of a camel you have to allow for how they sway from side to side." The policeman glared at him, but Alex kept a straight face, so finally the officer said, "Does anyone else in this car have a driver's license?" Someone else showed his license, and the officer said, "*You* don't come from Egypt, do you? Okay, then change seats with him."

About 37 years later, I was in Egypt at a meeting of the International Statistical Institute listening to a lecture in a darkened room and someone came and sat beside me and whispered, "Hello, George." It was Alex. He told me he had given up statistics and was growing oranges!

In North Carolina, I took some driving lessons from a woman instructor who told me I must be "slow with the hands and quick with the feet." Later I had to have a licensed driver with me, and Stu Hunter bravely undertook the task. I think Stu became rather fed up with this, because before very long, he suggested that I take the driver's test. At that time in North Carolina, the police did the testing. I wasn't much of a driver, and on the appointed day, I was very nervous and did everything wrong. Stu witnessed my performance during the test and expressed his surprise when the policeman started to fill out a form saying I'd passed. Somewhat defensively the officer said to Stu, "Well he didn't actually *break* the law."

On another occasion, when I was still a novice driver, I foolishly shot out onto a main road without stopping. A motorist who had the right of way on the main road narrowly missed me and came to a halt on the opposite side of the road. He rolled down his window. I was expecting what would have happened in England at this point: a lot of bad language. But all the driver said was, "Well, hi!"

Later, I got a flat tire one Sunday morning. A car came along with a man and his family, who were in their best clothes, and I suspect they were on their way to church. The car stopped, and the man asked what seemed to be the trouble. I was embarrassed to tell him I didn't know how to change a tire, and I asked if he could perhaps tell me how to do it. He said, "I'll change the tire for you. You watch me and you'll know what to do next time." He got down on his hands and knees in his Sunday suit and changed my tire. I was so overwhelmed by his kindness that I didn't know how to thank him. He said, "Well you know we're all here to help each other," and he drove away. On my return to England, when I sometimes encountered what I thought were unjust criticisms of Americans, I told these stories.

Some years later, I needed to change a tire on my car. My son Harry, who was then five years old, enthusiastically helped me change the tire, and when it was finished, he said, "Now let's change the other three."

My year in Raleigh gave me a view of American social habits that were unfamiliar to me. Initially, my only obligation was to help Stu Hunter get his Ph.D., and during the year, I did the same thing for another student, Sigurd Anderson. Sig had recently married Sally, and Stu was married to Tady, and we had a fine time together. It started off with what my American friends called a "Round Robin Dinner" in which a number of graduate students took part. The idea was that the first course was prepared and served at somebody's house, the next course at someone else's house, and so on. It turned out that Jessie and I were responsible for the pre-dinner drinks. We had no experience in this. We knew that people in America drank cocktails because we'd seen it in the movies, but we knew nothing about mixing drinks except that they had to be very strong. What we served must have been close to neat gin, I think, because one of the ladies passed out before the first course.

Toward the end of my stay in Raleigh, Miss Cox told me we should take four or five weeks off to see something of the United States. We began by attending a conference in Canada, where a number of statisticians took the Thousand Islands boat trip. We needed gas at one point and stopped on the U.S. side. Immediately someone from U.S. immigration jumped on board and started asking us where we were from. Sigurd Andersen replied, "Denmark." I said, "England," and so it went. Everyone on the boat was from a different country. Finally, when Wassily Hoeffding said, in his very thick accent, "Russ-i-a," the man shouted, "Okay, that's it! Get this boat out of here!" So went the Cold War. (I can't remember what we did for gas.)

I remembered that I had two cousins living somewhere near Chicago whom I had never met. They were the son and daughter of my Uncle Pelham. I had no idea of how I might find them so I wrote to the mayor of Chicago. Very shortly I received a reply as follows:

116 Form M.O. 3 5M 11-51

MARTIN H. KENNELLY
MAYOR

CITY OF CHICAGO
OFFICE OF THE MAYOR

January 22, 1953

Mr. G. E. P. Box,
2706 Cartier Drive,
Raleigh, North Carolina

Dear Mr. Box:

    Your recent letter was received requesting assistance in locating your relative, Mr. Lester Box.

    We are turning your request over to our Police Department, for attention, this being the only agency available for assisting in such cases. We hope they can be helpful to you in this matter.

Sincerely,

B. C. O'Neill
Secretary to the Mayor

cc: Police Department

When Jessie and I drove from North Carolina to Chicago, my driving was far from perfect, especially in congested cities. I ended up in the Chicago Loop at rush hour. I wanted to turn right, but whenever the light changed, a mass of people started crossing and I didn't see how I could turn without running them all over. Finally a policeman put his head in my window and said, "Say, buster, don't we have any colors you like?"

My cousin Evelyn was married to a lawyer who was on the Chicago city council. He was very kind and took Jessie and me to a council meeting to see what it was like. My other cousin, Lester, worked in the car industry about 20 miles north of Chicago. He had a very lively family, and while we were visiting him, we went out to go dancing at about 11 o'clock in the evening, which was rather late for us.

While in Chicago we visited an English friend, Professor Brownlee, at the University of Chicago. He was tremendously enthusiastic about the beauties of the United States, and when we said we weren't sure where to go on our trip, he invited us over to a spectacular slide show in his office. He had hundreds of slides that he had taken of his many travels, and he gave us an illustrated commentary with a map. He was particularly intrigued, and so were we, by the meandering of the Colorado River, the Goosenecks of the San Juan River in Utah, the Black Canyon of Gunnison National Park in Colorado, the Painted Desert in Northern Arizona, and of course, the Grand Canyon. We decided to follow the route that he sketched out.

After we left the traffic of Chicago, our journey to the West progressed smoothly until we were about 200 miles from the Grand Canyon. There, the car that Alex had so methodically helped me choose broke down. It took two days to fix, and the fix, unfortunately, was short-lived, for as we neared the Canyon, it broke down again. We hung about while it was being repaired and began to wonder whether this part of the trip was worth the bother. There was a cowboy-looking guy leaning on a post nearby, so I asked him, "This Grand Canyon—what's it like?" He thought for a minute, and then he said, "Well, I'll tell ya, it's just about the biggest hole in the ground you ever did see!" We went, and it was. It was also one of the most beautiful and awe inspiring sights we had ever seen.

On the trip we drove through country where there were many Native Americans living in tepees. Our friend had told us that if we wanted a chance to talk with them, we must bring a gift—an apple would do. So we stopped at a tepee and someone came out, took our apple, and then disappeared into the tepee never to reemerge.

Our journey required that we travel a long way through the desert, so we had to have a carboy full of water before we started. So much of the West was remarkable to us, but one of the most interesting sites was an Anasazi Indian village at Mesa Verde where dwellings had been cut out of the side of a cliff. These had been abandoned many centuries ago, and there were signs that the inhabitants had left in a hurry. The climate was so arid that an old woman had been found in a remarkable state of preservation after 500 years. When we went to an evening campfire show put on by the Park Service, we learned more about the culture of the Native Americans who had lived in the area. Descendents of the Anasazi danced for us and told us about the system of lookouts that had warned their people of danger—most likely to come from their enemies, the Utes. They demonstrated the call that told the cliff dwellers that "the

FIGURE 5.3
Jessie and our first English car (a Hillman).

Utes are coming!" although they allowed that this was sometimes used just to keep people on their toes.

Jessie and I very much enjoyed our trip around the United States. One day, when we were on a narrow road in the mountains, we met a massive flock of sheep that formed an impassable barrier across the road, so while I drove the car slowly forward, Jessie sat on top of the car, banging on the metal hood and shouting loudly at the sheep so that they made way for us. The Native American sheepherder looked at me appreciatively and said, "You have good woman there!"

We carried memories of this trip when we returned to England, this time on the French ship, the *Île de France*. On this occasion, we had a pleasant trip. I remember there were a number of French Air Force cadets returning home. One group amused itself by dropping puff balls on the dancers, testing their bombing skills by aiming for the rather expansive cleavages of the ladies below:

Upon our return to England, we bought a new car—a Hillman—but we never forgot the one that Alex helped us buy.

Chapter Six

Princeton

$\mathcal{W}$HILE I was at the University of North Carolina, I was invited to give seminars at various institutions. One of these was Princeton where I met John W. Tukey, an extremely able mathematician and statistician. In addition to his work at Princeton, John had an important job at Bell Labs where he practiced and encouraged the use of statistics. John and I respected one another, but we didn't always see eye to eye.

Some of my early research at ICI had concerned "tests of statistical significance." That an effect is "statistically significant" means that it is unlikely to be a result of chance. You might, for example, be testing the efficacy of a new drug and you might want to check that the difference in efficacy between this and the standard drug was not just a result of experimental error.[1] Clearly questions of statistical significance must be considered because without them the scientist can, on the one hand, be "chasing red herrings" or, on the other, be missing important small differences.

Now there are a variety of tests of significance to choose from and they all involve assumptions. In particular, we may assume that we know the form of the probability distribution of the data (the probability distribution of the "noise"). Some tests assume that this distribution is of a particular kind called the "normal" distribution and this is in fact a distribution that quite often can approximate reality. The difficulty is that although while some important statistical tests are insensitive to large departures from important assumptions like normality, others are not. In 1953, I decided to call tests that are insensitive to a particular

---

[1] Notice that to say that a result is statistically significant does not mean that it is important.

An Accidental Statistician: The Life and Memories of George E.P. Box, First Edition. George E.P. Box.
© 2013 John Wiley & Sons, Inc. Published 2013 by John Wiley & Sons, Inc.

assumption "robust" tests, and this has become a generally accepted term. It is important to understand that it is the nature of the *test* as well as the nature of the assumptions that affect robustness. For example, if you ask the question, "Is A bigger than B?" an assumption of normality of the distribution might not be important, but if the question is, "Is A more variable than B?" it could matter a lot. One of the problems I worked on, therefore, was how to produce tests that are robust to assumptions. A robust statistical procedure, although not necessarily "optimal" for any *particular* set of experimental conditions, would work well in practice over a wide, relevant range of conditions likely to be met in practice.

Among his many abilities, John Tukey had a flare for coming up with ad hoc statistics that were "robust" to particular contingencies. Bob Hogg, and many other statisticians, did likewise, so there was a plethora of robust statistics. I preferred to achieve robustness using a Bayesian approach by introducing a suitable model that allowed efficiently for specific likely contingencies. These differences of approach led to my writing the song, "Let's Go Robust," based on Cole Porter's "Let's do It":

John does it
Hogg does it
Every statistician that's in vogue does it
Let's do it
Let's go robust

Outliers
Bad data,
Totally excluded by this estimator
Let's do it
Let's go robust

I got this
De-scending Psi function
That is free from all flaws
Moreover my function
Is much better than yours

A students
C students
Now all get robustified to B students

Let's do it
Let's go robust

We had our
Nice observations
Circumcised from both ends[2]
Removing the data
On which our theory depends

We tackle
Difficult cases
We can hit an outlier at fifty paces
Join the movement
Let's go robust

Box only
Does it with models
In all kinds of different ways
Claiming he learned it
From some preacher called Bayes.

In fame and fortune
You can be sharing
If you find a way to do it that's a wee bit daring
Strike out now
And go robust

Looking at the problem in the light of today's technology, you can see that the basic problems of tests of significance and robustness were solved in principle at very early stages of Fisher's thinking. From the first Fisher insisted on randomization to avoid biased conclusions, but he also pointed out that randomization itself provided nonparametric significance tests. One calculated what had happened with all the values of the relevant criterion that *could* have happened if the randomization process had occurred differently. These tests were not used at the time because of the lack of computing power, but are now possible with the advent of fast computers.

---

[2]This is a reference to a procedure called "Winsorizing" in which, in an endeavor to remove "outliers," the largest and smallest observations are omitted.

After I returned to England from North Carolina, I kept getting calls from John. He wanted me to leave ICI and come to Princeton to be the director of the Statistical Techniques Research Group (STRG) that was being set up there. Finally, in 1956, Jessie and I agreed to go along with our newly adopted son, Simon. John had arranged housing for us, a log cabin in the woods. When we arrived at the house for the first time, there was a man there who said, rather bluntly, "I'd like to see the money." After we had been in Princeton for a few months, we were dining with the Hunters when Jessie announced that she had befriended "a wonderful kitty" at the cabin. As she continued her description of this animal, Stu and his wife realized that the "kitty" was in fact a skunk (a creature that doesn't exist in England). After living at the cabin for a year, we moved to university housing where the wildlife was more familiar.

During the years I spent at Princeton, the STRG did some excellent research, producing many publications. Members of the group, many of whom came as visitors, included Stu Hunter, Don Behnken, Colin Mallows, Geoff Watson, Merve Muller, Norman Draper, Henry Scheffé, Martin Beale, and H. L. "Curly" Lucas. Henry, who was from Columbia University, was an important visitor. He was concerned with our physical, as well as our mental, well-being—he insisted that we all go swimming every lunch time and led us to the pool. Stu, who was a permanent member, recalls driving John Tukey's rusty station wagon to meet Colin Mallows at the boat in New York City, and in a moment of excitement, Colin put his foot through the floor of the car.

When we first set up the group, it was housed in the Theobald Smith House on the Forrestal Campus. This address foxed some of our correspondents. In particular, we received a request for a report that was addressed to "The Old Bald Smith." I found a photograph of a distinguished looking elderly bald man, and we framed it, hung it on the wall, and claimed that "Old Bald Smith" was our founder. You need tradition to be respectable.

Later the group moved to two houses on Nassau Street that were to be made into one. When I signed off on the plans before I visited England, I noticed that one thing to be done was the removal of a first floor closet. Upon my return, I noticed that the closet was still there, and when I pointed this out, I was told that they had kept the closet because it was the only thing that was holding up the bathroom on the second floor.

Much as Francis Scott Key had been inspired by the events of his time, the closet prompted Collin Mallows and I to write the following ditty:

If you're walking by the Gauss House
And look inside the door,
You will see a little closet
And you'll ask "Now what's that for?"
All the members of the group will come hurrying along,
They'll stand round in a Normal curve
And sing this little song:
"It's the closet that holds up our john,
The basis that all rests upon.
Without it the structure would all tumble down,
It's the closet that holds up our john!

We called the building on Nassau Street the "Gauss House" because of our admiration for the great mathematician, Carl Friedrich Gauss, and we nailed up an appropriate sign. But this led to a misunderstanding because there had been a famous Princeton dean, *Christian* Gauss. One day it was raining hard and on leaving our building Merv Muller encountered an elderly lady sheltered by an umbrella who was staring at our sign. She said, "Dean Gauss never lived in that house!" Despite the fact that he had no umbrella and was getting very wet, Merve explained the misunderstanding and described at some length the many accomplishments of Carl Friedrich Gauss. Unmoved, the lady exclaimed, "I don't care what you say; Dean Gauss *never* lived in that house!"

In the mid-1950s, one of the topics of interest was the modeling of chemical reaction mechanisms described by differential equations. This required nonlinear estimation and a great deal of computing. I got together with a professor of engineering, and we found that by pooling some of our research funds, we could rent an IBM 650 machine. Judged by today's standards, this machine was painfully slow and had very limited memory, but at that time, it was the last word in computers. We needed to get permission from the Dean to use our money for this purpose. He said that he wouldn't try to stop us, but that we were being very foolish. He explained that they had made a survey of all the people interested in computing at Princeton and had found that with such a "very powerful machine," all the University's needs for a whole year would be satisfied in one afternoon of computing time! As we suspected, this was a mistaken view. Within a few months' time, our rented machine was used 24 hours a day. Our group used it during the day and made it available free to anyone who wanted to use it at any other time.

I had attended my first Gordon Research Conference on Statistics in 1953, while I was at North Carolina. The Gordon Research Conference was named for Professor Neil Gordon, a chemist at Johns Hopkins who had become discontented with the habit that "big shot" researchers had of presenting their papers at scientific meetings and then disappearing. He felt that opportunity for informal discussion with colleagues was essential. In the 1930s, the conference was held on Gibson Island, in Chesapeake Bay, and all the participants were marooned there for a week. However, the idea of the conference soon spread to other topics and locations. Thus, the Gordon Research Conferences on Statistics in Chemistry and Chemical Engineering began in 1951 and continued through the summer of 2005. For many of those years, it was held on the campus of New Hampton School, a private boarding school that was beautifully situated in a small New Hampshire village.

The conference lasted five days, and statisticians, chemists, and chemical engineers came from all over the country. Many flew into Boston where there was a bus waiting to drive them to New Hampshire. Don Behnken and I often drove together. He later recalled our "automobile trips to the Gordon Conference, forking left and forking right with an average of a punch line or a song every two miles [and] some memorable walks and conversations at the same conference."[3]

I had become close friends with Don and his wife, Neeta, while I was at Princeton. Don had served in the U.S. Navy from 1943 to 1946. He had undergraduate degrees in physics from Dartmouth and Yale and an MBA from Columbia. In 1956 he had enrolled in the Ph.D. program at Raleigh. His friend, Sig Andersen, urged him to study with me, but by then I was at Princeton. Undeterred, Don spent two summers working with me at Princeton on the understanding that he would qualify for his Ph.D. at Raleigh. In 1960, we produced what came to be called Box–Behnken designs, which are experimental designs

---

[3] Donald Behnken, letter to author, April 6, 1984.

for response surface methodology.[4] These use only three levels for each variable and allow the fitting of important models with few experimental points.

After Don received his Ph.D., he went to work as a statistician for American Cyanamid in Norwalk, Connecticut. Later I became a consultant for Don's company, which gave me a chance to work with him and visit his family. Don was a wonderful sailor, and I spent a many weekends on his boat. Once, when his family was not with us, Don and I passed the weekend sailing and telling jokes. Behnken claimed 106 were mine and 30 were his.

On another occasion, when his family was on the boat with us, we took a picnic to an island in Long Island Sound. Soon after we had arrived, a serious storm blew up and we rushed to get ourselves and the picnic things into the dinghy and back to the anchored boat. It was a matter of minutes before the dinghy swamped, and the adults and children all set to bailing furiously as we fought through the waves and just managed to reach safety.

Don and I continued both the banter and more serious shoptalk at the Gordon conferences. As Neil Gordon had intended, the conference provided free time for informal exchanges of this kind. Although mornings and evenings were reserved for formal presentations and discussion, the afternoons were open. There were many beautiful places for walking, and close by was a place to swim. Spouses (in those days mostly wives) were invited, and there were activities specifically for them. The food was particularly good, with lobster being the traditional fare of the final evening. On the last night there was always a skit. Below is an excerpt from one in which I played an aberrated version of myself.

The scene is the office of the senior professor George, at the University of Little Marsh in Radison. The faculty, who are not well endowed, have pulled their last resources together to send a junior faculty member, young Professor John, off to the Gordon Research Conference to learn about new things. We see young Professor John come into the office of Professor George.

George: Who are you young man? I don't recognize you!
John: I am an assistant professor.

---

[4]G.E.P. Box and D.W. Behnken, "Some New Three Level Designs for the Study of Quantitative Variables," *Technometrics*, Vol. 2, No. 4, 1960, pp. 455–475.

George:	Which department?
John:	Your department, Sir!
George:	Ah well, okay, what do you want young man? Asking for a raise I suppose . . .
John:	Well, we agreed that we meet here and I report about the Gordon Research Conference. Will the Dean come around?
George:	Yes, he'll come a bit later. He is messing around somewhere.
John:	This conference was most confusing! First I thought I had come to the wrong conference, that one on zoology. They only talked about milking fish patties and on how to convert pesticides into hair growth agents. But after a while they went into more specifics and taught us how to run experiments.
George:	But we know that, don't we?
John:	Yes but they had some strange ideas about experimentation. Instead of changing one factor at a time, they change them all together!
George:	Most unscientific! That is ridiculous! You all remember my famous experiments supported by the Southern Neural Fundamentalists Network (SNAFU), in which I refuted once and for all this theory of sexual reproduction, the details of which I won't even repeat here—it's too disgusting. For these experiments I used two rabbits, a doe and a buck. They said that a doe and a buck would produce little rabbits, but I proved them wrong.
	Here Professor George shows a series of overhead projections.
George: *continues*:	Thus they were forced to accept the alternative hypothesis that little rabbits are brought by little storks and hidden under gooseberry bushes . . . Oh, here comes our queer old Dean. We are listening to this young man—what did you say your name was?—reporting from the Gordon Research Conference.
Dean:	Oh yes, here you are, young John!
George:	They had some very strange ideas of changing all experimental factors together and I reminded them about our rabbit reproduction research.
Dean:	Yes, and your famous flea research!

George:	Yes, again I've proven them wrong! I took a flea. I banged the table and it jumped. I banged the table again, and it jumped again.
John:	That's a replication.
George:	Nonsense! I just did it again. Well, I cut the legs off the fleas and I banged the table and it did not jump. I banged the table again and it did not jump.
John:	That's replication again!
George:	You hold your tongue, young man! What I proved was that when you remove the legs of a flea, it loses its sense of hearing!

And so on.

One very interesting man who regularly attended the Gordon Conferences was Frank Wilcoxon (Figure 6.1). I looked forward to seeing him and his wife Freddie on these occasions. In particular, I fondly recall that they would often invite me for an aperitif before dinner.

Born in 1892, Frank had had an adventurous youth before receiving his Ph.D. in physical chemistry at the age of 32. In the mid-1920s, he had a postdoctoral fellowship at the Boyce Thompson Institute for Plant Research where he worked on the use of copper compounds in fungicides. It was at the Institute that he and his colleagues, among them Jack Youden, studied Fisher's *Statistical Methods for Research Workers*, published in 1925.[5]

In 1943 Frank began to work for American Cyanamid, where he led a group at the insecticide and fungicide laboratory. It was at American Cyanamid that Frank's main interest moved from chemistry to statistics, and in 1945, he wrote a paper that shook the field.[6] The background to this was that Frank's lab was required to test a large number of synthetic substances, comparing them for effectiveness with the previously best-known product. Frank had to check all these comparisons for statistical significance, and there were a very large number of these to do. The

---

[5] R.A. Fisher, *Statistical Methods for Research Workers*, Oliver & Boyd, Edinburgh, 1925.

[6] F. Wilcoxon, "Individual Comparisons by Ranking Methods", *Biometrics Bulletin*, Vol 1, 1945, pp. 80–83. Reprinted in the Bobbs-Merrill Reprint Series in the Social Sciences, S-541. Also contained in the pamphlet by Frank Wilcoxon and Roberta A. Wilcox, *Some Rapid Approximate Statistical Procedures*, Insecticide and Fungicide Section, Stamford Research Laboratories, American Cyanamid Co., 1964.

FIGURE 6.1
Dr. Frank Wilcoxon.

standard statistical test was Gosset's *t*-test, and unfortunately, with the primitive calculators of the 1940s, this took a long time. So Frank invented a quicker test. To determine the level of significance you in effect simply noted the overlap between the two samples and consulted a table that Frank had calculated.

In 1950, Frank began work at the Lederle Laboratories Division at American Cyanamid. There he began a statistical consulting group that he ran until he retired in 1957. In 1960, he took on a half-time distinguished lectureship at the newly formed Department of Statistics at Florida State University and taught until he died on November 18, 1965.In January 1967, I joined others who gave lectures at a conference in Tallahassee in Frank's memory. Franks wife, Freddie, was there, as were numerous colleagues. Frank and Freddie had been avid bicyclists, but as he aged, Frank began to ride a motorcycle, which you can see in the photo in Figure 6.1.

Cuthbert Daniel also attended the Gordon Conference regularly. He was born a maverick in 1904 and remained one his entire life. He was a chemical engineer who had never attended a course on statistics, but who had read Fisher's *Statistical Methods for Research Workers*, which his wife Janet had brought home as part of her studies in biochemistry. Cuthbert was among the first to employ factorial experiments in industry. He invented a simple graphical method for analyzing data from such designed experiments. His method also detected suspect data values. From the 1940s through the mid-1980s, he worked as an industrial statistical consultant for Union Carbide, Proctor and Gamble, U.S. Steel, E.R. Squibb, and others and regularly came up with highly original and useful ideas.

Cuthbert did not suffer fools gladly and had a hearing aid with a cord and a switch that he sometimes rather pointedly turned off. He had a remarkable sense of humor and drily pretended to be overawed by theoreticians. He was liable to say, in the middle of his own talk, "Alright *so* far!" I remember once, during a very mediocre presentation that we had to listen to, Cuthbert whispered to me, "George, there's less in this than meets the eye."

On another occasion on the lawn after a meeting, I was asked by someone about the method of least squares. I said, "Well, you see you have a vector x representing the input variables," and I moved my left hand to point to the direction of this imaginary vector, "and another vector y representing the data," and I moved my other hand to point to the direction of the output vector. At that point, Cuthbert, as he was walking past, said, "Let me hold that one for you George."

Cuthbert could be dramatically informal. One evening when we were consulting in Pittsburgh, he suggested going to see a play in which his sister was appearing. It was some distance away, and I forget how we got there, but at the end of the play, it was far from clear how we were going to get back. Cuthbert solved the problem by saying to two people in a car, "Are you going to Pittsburgh?" When they said, "Yes," he got into the back of their car, opened the door for me to do likewise, and told them where he wanted to go. I think the middle-aged man and his wife were terrified. They must have thought that we were gangsters, but they took us there. Cuthbert thanked them, and we got out.

I have in my possession some wonderful letters that Cuthbert wrote to me in the mid-1950s, some addressed "Dear E.P." One begins:

> Dear George,
> Apparently even a semi-annual letter to you strains my capacities. But here I am. It is a sticky day and so I am boycotting the national convention of ASQC and catching up on my home work. I want you to answer this half-normal plot thing. Please don't think you can get rid of me, or even change our friendship by casting me into outer darkness, saying—Ye are a Fisherian, or something repulsive like that.

Cuthbert remained my friend until he died in August 1997 at the age of 92.

For many years, Stu Hunter and I were frequently on the road together, often as industrial consultants for the same company, and when we taught week-long short courses for industry in various places in the United States and Europe. It was at a Gordon Conference that Stu and I met Dr. Frank Riordan. Frank, who was a chemical engineer at Chemstrand, was faced with a very difficult challenge. The DuPont Company had previously had a monopoly on the manufacture of nylon and had consequently been sued for constraint of trade. After many years of litigation, the following solution was agreed upon in 1951: DuPont would build a second nylon plant at Pensacola, Florida, using all their expertise, with a capacity to match the DuPont plant.

Frank was an engineer responsible for process improvement in the new company. Since DuPont had been making nylon for many years, he knew that to compete with them would be tough. He believed, however, that by using data analysis and statistical experimental design to solve problems, he could make up the difference. So he invited Stu Hunter and me to come on board as consultants.

When the new plant was built, it was immense, extending over 2,000 acres. The plan was that DuPont would start up the new plant and then, on a certain date, they would hand it over to the new company, which would be called Chemstrand and would be affiliated with Monsanto. After this, DuPont and Chemstrand would be competitors. One story that went the rounds was told by the new plant superintendant. He was looking over the plant a few days before production was scheduled to start when he met a process worker who asked him who he was. He said, "I'm the plant superintendant." The process worker looked at the huge plant that stretched away in the distance as far as the eye could see and said thoughtfully, "Jeez, if you don't screw up you've *really* got it made."

Frank had a small but impressive technical staff of chemists and engineers, and for several years, we met with him and his group regularly. Together we solved many problems, usually with the help of carefully planned experiments. I remember one problem very vividly. At one point Chemstrand's nylon cord, used in the manufacture of automobile tires, was not as good as DuPont's. Salesmen complained, and we feared we would lose this market to DuPont. No one in the company wanted to tackle the problem, so it was left to Frank's group. Frank put a sign on his door that said, "Calhoun." The joke was about a boys' school football team that got involved in an exceedingly rough game. The coach kept shouting, "Give the ball to Calhoun! Give the ball to Calhoun!" Eventually a call came back, "Calhoun says he don't want the ball!"

We made numerous trips to Pensacola, which, in the 1950s, was largely pine forests and undeveloped coastline. During our free time, we enjoyed the Gulf of Mexico, which offered a beautiful place to swim and eat fresh seafood. Frank was a great cook and a jazz aficionado, and my consulting visits were enlivened by jazz band parties at his house, which were attended by friends who played various instruments. I did my best to keep up on the guitar.

Stu and I had learned that sometimes things on the road went smoothly and sometimes they did not. A plane that was supposed to fly sometimes didn't fly, and once, when I thought I was taking a plane to Decatur, Alabama, I ended up in Decatur, Georgia. Stu once told me a story that helped when things did not go according to plan. It went like this: A couple had twin boys. One was an incurable pessimist, while the other was a total optimist. On their birthday, the parents resolved to cure them of these peculiarities. They took the pessimist to a room with every kind of present a boy could want. They took the optimist to a deserted stable that contained nothing but horse manure and told him that his present was in there. After about an hour, they checked on the boys. They found the pessimist in tears. He didn't understand the instructions for his toys, and some he had broken. When the couple looked for the optimist, they found him wandering around the stable with a happy smile on his face, saying, "There's *got* to be a pony in here some place."

That phrase sometimes came in handy.

In the late 1950s, Stu Hunter, Cuthbert Daniel, and I decided that a new journal was needed to meet the needs of those who applied statistics in technology (Figure 6.2). The first half of the twentieth century had shown the value of statistics to scientific research, thanks largely to Fisher and Walter Shewhart. The momentum continued with the beginning of the Gordon Research Conference on Statistics in Chemistry and Chemical Engineering in 1951, and the start of the American Society for Quality Control (ASQC) in 1952. Applied statisticians were still a

FIGURE 6.2
Stu Hunter, me, and Cuthbert Daniel.

rare breed, but with the postwar growth in industry, there was growing interest in the application of statistics to technology.

In 1957, some months after I had arrived in Princeton, Professor John Whitwell of Princeton's Chemical Engineering Department invited me to give a talk on iterative experimentation at the Gordon Conference, which he was chairing that year. At the conference, there was informal discussion about the need for a new journal. Later that year, Horace Andrews, who taught statistics at Rutgers and who also headed the Education Committee of the Chemical Division of the ASQC, asked me to develop a course on statistical methods for process development. Stu Hunter, Cuthbert Daniel, and I planned the course together and once again discussed the prospective journal. We broached the subject with members of the ASQC in 1958, and out of our discussions came a prospectus calling for a "new journal to present applications of statistics in the physical and engineering sciences."

We decided that the journal should be a joint project between the ASQC and the American Statistical Association (ASA), with the governing body containing a number of members from each institution. The idea was that the ASQC would ensure that the new journal kept its feet on the ground while the ASA would guarantee high quality.

To start the journal, we needed about $10,000. The ASQC immediately came up with $5,000, but the ASA was hesitant to provide funds. So Stu, Cuthbert, and I put on a special course. We raised the $5,000 and gave it to the ASA.

The first issue came out in 1959 and bore the subtitle "A Journal of Experimentation in the Chemical and Other Physical and Engineering Sciences." Stu was the first editor, and R. A. Fisher, who suggested the name *Technometrics*, contributed a paper, "Mathematical Probability in the Natural Sciences."[7]

I had no teaching duties at Princeton, but in 1959, the Chemical Engineering Department asked me, as a favor, to teach a course on experimental design. They said that the class would be for graduate students and that I did not have to give any exams or grades. Despite this, at the end of the semester, I got a phone call asking what grade I had

---

[7] See D.M. Steinberg and S. Bisgaard, "Technometrics: How It All Started," redOrbit.com, March 10, 2008.

given to a student called William Gordon Hunter. I said, "I thought you said I didn't need to give any grades." They then told me that Bill Hunter was not a graduate student but an undergraduate in chemical engineering with special permission to take the graduate course. (Bill later told me that in order to take the class he had had to seek authorization from five deans!)

I was perplexed about what to do, so they said I could give him an oral exam. Up to that time, I had not taken much notice of Bill, but at the oral exam, he really surprised me with his knowledge. Princeton had a grading system of 1 through 7, with 7 being the highest grade. When I informed the office that Bill had received a 7, they said that was good because he would now graduate *summa cum laude*. Bill had told me that he planned to go to Illinois for a year to get a Master's degree in chemical engineering. After that, he wanted to do a Ph.D. with me at the University of Wisconsin.

Chapter Seven

A New Life in Madison

$\mathcal{J}$ enjoyed my stay at Princeton and was to receive a full professorship there. But sadly in 1959, I had gone through a divorce, and Jessie and Simon had returned to England. I was to marry Joan Fisher, and to spare her some of the inevitable gossip, I decided to leave Princeton. Sam Wilkes, the head of the department, very much wanted me to stay at Princeton. He argued that in a year or two, this would all be forgotten. I felt adamant about leaving, however, and sought a job elsewhere. I found out that Columbia, Chicago, and Berkeley were interested, but at Wisconsin, there was the attractive possibility of starting a new department.

So what else happened so that I ended up in Madison? I think it went like this: Henry Scheffé, with whom I had worked at STRG, was on the faculty of the Statistics Department at Berkeley. He believed that to be a good statistician, you needed more than good mathematics. Henry had read my papers and gave a seminar at Berkeley called, "Some Bright Ideas of G.E.P. Box." Meanwhile, Professor Rudolph Langer, mathematician and director of the Mathematics Research Center (MRC) at Wisconsin, had been looking for talent for the MRC. Langer, whose search was for statisticians as well as for mathematicians, knew and respected Henry, and I believe it was Henry who first told Langer about me.

At Madison there was a loosely associated group of perhaps 200 people called the "Division of Statistics." All you needed to be a member of this group was to have some interest in the subject of statistics. The group's members had decided that it wanted to set up a statistics department at

An Accidental Statistician: The Life and Memories of George E.P. Box, First Edition. George E.P. Box.
© 2013 John Wiley & Sons, Inc. Published 2013 by John Wiley & Sons, Inc.

Madison, and the idea was that if I came and formed a department, I could begin by being employed part time at the MRC.

I was invited to give two seminars, one on some technical problem and the other on what I would do if I were asked to start a new department. My plan was to have a central statistics group, plus joint appointments in agriculture, engineering, medicine, and business. The next day the dean told me, "The committee liked your ideas and wants you to come and put them into practice."

What was remarkable was that I was appointed to initiate and head the department as a full professor even though I had never had an academic appointment at any university. At North Carolina, I was called a "visiting research professor." At Princeton, I had been a "senior research associate."

The new department was not to be started until the beginning of the academic year in September 1960. In the months that intervened, I worked at the Math Research Center and planned the new department. We needed additional funding, and Dean Erwin Gaumnitz of the Business School was instrumental in helping the department get its financial legs. The accommodation for the new statistics department was modest at first. We were housed in a Quonset hut, which was one of many erected to provide housing for the influx of returning soldiers who enrolled at the University after World War II. The hut was near the lake, and from time to time, it flooded. At first there were just me and a secretary occupying the hut and getting wet.

The Math Department at Madison was anxious to get rid of the statistics courses that it had offered in the past. I had little formal teaching experience when I came to Madison, but there were parallels between what I had to do in the classroom and what I had done in the Army and at ICI. There I had worked closely with scientists to use statistics to help solve problems and I had given occasional talks about problem solving. A flier would be posted and a small group—perhaps six or seven people—showed up to listen. I also taught a night class at Salford Technical College, halfway between Blakely and Sale, where I lived. I suppose events of this kind prepared me, to some extent, to teach in a university classroom.

Once I got to Madison, it was my first duty to teach the course the "Advanced Theory of Statistics." I issued mimeographed notes each week, and as a student of Egon Pearson, initially I taught Neyman—Pearson theory. However, after I left University College, partly because of research I had done on what happened when standard assumptions were *not* true, and partly from what I thought was common sense, I had found a Bayesian approach much more convincing. So I found that my teaching of this course became more and more Bayesian with every week that passed.

From the beginning, I wanted my students to know that they were at the forefront of a movement in which statistics played a vital role in *scientific inquiry*. I wanted them to understand that statistics was a catalyst to learning and discovery that had many useful applications in the science and engineering fields. Moreover I wanted them to take their ideas out of the classroom, to discuss and to argue them, and to meet industrial statisticians who could explain how *they* solved problems.

No beginning teacher could have had a more promising group of students. George Tiao, Bill Hunter, and Sam Wu were among my first students, and by 1963, all three had taken their PhDs at Wisconsin. Dean Wichern, who came later in the 1960s, became a professor in the business school in 1969. Other students from that first decade, such as Duane Meeter, David Bacon, and Paul Newbold, went on to teach at universities in the United States., Canada, and England. John Wetz, Bill Hill, David Pierce and Jake Sredni found employment in industry or government. I have said many times that as a student, George Tiao was my bell weather: Whenever he looked worried, I examined the blackboard to see what I had done wrong.

I did not realize that when David Bacon was my student, he enjoyed scribbling some of my odd remarks and unusual turns of phrase in the margins of his lecture notes. He recently sent me some of these, which he has preserved since the 1960s:

Fisher is being a bit "shirty" here [(in specifically denying the legitimacy of assuming locally uniform prior distributions in the absence of prior knowledge).]

Likelihood methods are like a very intelligent but nondiscriminating child.

Good research these days is simply a moving of one blinker three degrees to the right.

There may be some forlorn and shipwrecked brother who is traveling at the same speed as I am.

None of this is a hanging matter.

Whenever we see virtue rewarded, we are completely surprised.

At this point, I don't want any chemical engineers to spit on the floor and walk out.

You think that because an F-test shows significance, you can draw conclusions about the contours? Not on your Nelly!

What we need at this point is a blackboard that can be erased!

[Said during a Monday night beer session . . .], I am so angry at the College of Letters and Science that I have forgotten what I was going to say.

Ah Jeffreys, poor Jeffreys! Can't we then have Lady Bella?

Data cannot always speak for themselves.

In model building, by and large, it's easier to get things into the act than out of the act—and people too, for that matter.

[Working his way through a rather detailed derivation . . .]Take care not to get your nose caught in the tram lines.

Now we move from the sublime to the gor blimey.

[As George is about to launch into a detailed derivation . . .] Now, like the beginning of the English radio program "Living with Mother," "Are you sitting comfortably? Then let us begin."

I had better say this again because one chap in the back row looks as if he's been hit over the head with a sledge hammer.

Most of the members of the Institute of Mathematical Statistics have never seen a model, don't want to see a model, and couldn't care less about data. If you want to see some of these people, they'll be here in a couple of weeks. Be kind to them. Remember that that they *are* human beings, you know.

Over the years, the department was situated in four locations. After the brief stint in the Quonset hut, we moved to a nice house on Johnson Street. I was sitting in my office one day when a man came in and started knocking on the wall. I asked him what he was doing, and he said he was checking where the joists were. When I asked, "Why are you doing that?" his response was, "Well, as you know, we are knocking down this house next week." I immediately called someone in the University, and I was told, "Oh no, Professor Box. That house *is* scheduled for eventual demolition, but this is not likely to happen for another year or two." With relief I put down the phone, but after about 15 minutes, it rang again. An agitated voice on the other end of the line said, "You were right! Can you start getting out this afternoon?"

Next we were accommodated over Tiedemann's Drug Store on University Avenue. It must have previously been a rooming house because it had a lot of bathrooms. (The baths turned out to be handy places for the graduate students to park their bicycles.) The room we used for seminars was an L-shaped room, useful when we had an uninteresting seminar speaker because one could take refuge around the corner.

Dean Young (later chancellor) came to see me when we were at Tiedemann's. I asked him how you got a new building. He said, "You get in the queue." I said, "How do you jump the queue?" He replied, "You get matching funds from somewhere." We applied for National Science Foundation money, and when they came on a tour of inspection, June Maxwell, our secretary, enlarged the pages of some of our reports and hung them on the wall in strategic places. To emphasize our need, she also arranged that bits of plaster fell from the wall as people passed by.

In 1964, the University formed a Computer Science Department. They didn't have a senior figure to lead the way, so they got Professor Steve Kleene of the Mathematics Department to be temporary chairman. Kleene was a brilliant mathematician and a world-famous logician, as well as a charming man. Steve and I met and found we shared the belief that statistics and computer science could interact and complement each other. So our pitch to various possible funding sources was based on this idea. We talked to a number of funding agencies, and we finally got some money from the National Science Foundation. (When we were talking to the granting agencies, I might have been too verbose, because Steve Kleene rather pointedly told me a story about two men listening to a speaker, and after about five minutes, one said to the other, "Let's give him twenty dollars." After ten minutes, he said, "Let's give him five dollars." After twenty minutes, he said, "Let's not give him anything at all." I took the hint.)

We planned a building in which statistics and computer science offices would alternate so as to maximize the chance that people from the two disciplines would talk and work with each other. Also, there was a single lounge where they might discuss problems over coffee. Unfortunately, time proved our plan a miserable failure: Statisticians and computer scientists were not interested in talking to each other. After a while, computer science had the rooms rearranged into two separate blocks. Computer science expanded at a much greater rate than statistics, and in

2003, the statisticians were evicted to their current quarters, offices in a building that had been the old hospital.

One day, soon after the Computer Science Department got started, the new chairman came to my office to ask me about the regulations for the master's and Ph.D. degrees. I said, "What regulations?" He explained that an inquiry had revealed that a new department had to prepare various documents setting out requirements for degrees and so forth, and these had to be approved by appropriate committees. I had not done any of these things, so I couldn't help him. I felt that I had been entrusted with starting a department of statistics and I had assumed that I could develop it my own way. My rules for the Ph.D. exam, for example, were based on those at the University of London, which simply said that a candidate should submit an original thesis and the examiners could then (1) pass the candidate; (2) submit the candidate to any examination or examinations, written or oral; or (3) fail the candidate. Later on the Ph.D. regulations in the Department of Statistics were changed and became a long rigmarole like those of other departments. I am not really sure, therefore, that some of our early Ph.D.s were legal, although we produced some fine statisticians.

Both Joan and I are British citizens so before leaving England to come to the United States in 1959, we obtained alien registration cards. This entitled us to many of the privileges of a U.S. citizen, but getting such a card was not easy. We had to fill in the largest form I've ever seen. You had to answer a great many questions, like, "Are you now, or have you ever been, a member of the New York Photographic Society?" (You were left to fantasize on what they did in the dark room.)

There were however some more serious complications. I was already in the United States, but Joan was to come somewhat later with our three-month-old baby, Helen, who had been born in England. Helen, of course, did not possess an alien registration card, so beginning in May 1960, five months before the baby's birth, I inquired whether my daughter would need a visa to enter the United States. This began considerable correspondence.

I first wrote to the American Council for Nationalities Service in New York and received a long reply dated May 16, 1960, the meaning of which I found impossible to fathom. This quoted State Department Regulation 22 C.F.R. Section 42.36. I later wrote a letter to the Department of State in Washington and received another lengthy letter, dated June 10. This said that regulation 22 C.F.R. Section 42.36 would *not* be applicable to the baby's situation. Finally, on February 2, 1961, I received a second letter from my first correspondent from New York that seemed also to contradict the previous one.

In the end, Joan flew to the United States. with the baby in a basket, and she passed through the immigration and customs process with no difficulty. Afterward I inquired how I could legalize her status in the United States. I received an alarming reply informing me that there was a good chance the baby was in the country illegally and that we would have to leave and then cross back over the border to establish legality. Luckily, after more bureaucratic exchange, I was eventually able to add Helen to my passport, finally putting an end to her unintended life of crime.

Many years later, I had a similarly vexing experience with U.S. immigration authorities. The American government had decided that all resident aliens needed to be fingerprinted. The nearest immigration office was in Milwaukee, 70 miles from Madison. I went there and waited for what seemed an interminable amount of time. Finally, when it was my turn, the employee who was attempting to take my fingerprints was having trouble and called in another person to help. They both twisted my fingers in every possible direction, but they were unable to obtain prints. I was asked to return to the office on another day, when another attempt would be made but upon my return, the same thing happened. After considerable mulling and muttering, it was determined that I have no fingerprints. Like my infant daughter, I had launched an inadvertent career of crime. I thought immediately of:

> Macavity's a Mystery Cat: he's called the Hidden Paw—
> For he's the master criminal who can defy the Law.
> He's the bafflement of Scotland Yard, the Flying Squad's
> despair:
> For when they reach the scene of crime—*Macavity's not there*!
>
> He's outwardly respectable. (They say he cheats at cards.)
> And his footprints are not found in any file of Scotland Yard's . . . [1]
>
> T.S. Eliot

---

[1] From the T.S. Eliot poem "Macavity: The Mystery Cat," in *Old Possum's Book of Practical Cats*, 1939.

FIGURE 7.1
Harry, Helen, and our dog Victor.

My son Harry was born in Madison on May 13, 1962, becoming the only U.S. citizen in the family (Figure 7.1). We lived close to Lake Mendota in Shorewood Hills, an attractive village in the middle of the city of Madison. As they grew up, the children went to Shorewood School, the public school nearby, where they had the advantage of meeting other children of diverse ethnic backgrounds. Close by was university housing for visiting faculty and graduate students who came from all over the world. The children of these families made Shorewood School as international a place as the university campus.

Madison is a city of lakes, and families take advantage of this whatever the season. When the children were young, I bought a small sailboat and Helen and Harry learned to be proficient sailors. Winter also invited us onto the lakes. Once when Helen was a toddler, she and I took a walk on frozen Lake Mendota. We passed many ice fishermen, and Helen peered into each of their buckets to see what they had caught. Whenever she saw an empty pail, she exclaimed rather loudly, "This one hasn't caught *anything*! And *this* one hasn't caught anything either!" I thought we'd get shot.

Joan attended the university where she studied in the English Department. She earned an honor's degree in English and a master's degree in

FIGURE 7.2
Joan getting her degree with Helen and Harry.

the History of Science (Figure 7.2). In 1978, she wrote a highly praised biography of her father, *R.A. Fisher: The Life of a Scientist*, which was published by Wiley.

Both Joan and I enjoyed Shakespeare, and in the summer we would drive with the children to the Shakespeare Festival at Stratford, Ontario. Joan missed England, and Stratford had a bit of England about it, with its parks and gardens and places that served tea. Many of the actors were British, chief among them Alec Guiness, who was there in the earliest days of the festival.

In the mid–1960s, I received an invitation to be a visiting professor at the Harvard Business School. We packed up our car, and with our two small children in the back, set out for Massachusetts. Midway on our journey, we stopped at a motel. Joan and I busied ourselves removing the luggage from the car. It hadn't been a minute when we heard a happy voice exclaiming from a distance, "I like it! I like it!" There was Helen, arms extended, flying down a very tall slide and into the motel swimming pool. She hadn't yet learned to swim, so Joan leapt into the pool fully clothed to rescue her. With half of us wet through, we entered the lobby and checked in.

Near Boston we rented a large, dark, and spooky old house with turrets and extensive grounds that belonged to a professor of French who was on leave. Stories about the place abounded: A former owner had once kept 50 cats there. And some years before, a number of valuable paintings stored in the barn had been stolen. The heating system, which relied on several thermostats, was incomprehensible; some rooms were icy, while others were tropical.

There were numerous trees on the property, and one autumn day, Joan raked the fallen leaves into an enormous pile and made a bonfire as she would have done in England. Within minutes, and to the delight of my small son, the fire brigade roared up and extinguished the burning pile.

As a small boy, Harry liked cars and trains and planes. One Christmas when he was three or four, we gave him a bright red pedal car. He liked it so much that we could not extricate him for hours and we had to bring him his breakfast in the car.

Later, when he was about seven, I happened to see an advertisement in the paper for a Lionel train set, so I went to look at it. It was a beauty with a locomotive that smoked; a number of cars, tracks, and switches; and a station. The man who was selling the set explained that he had bought it for his daughter but she wasn't interested. So I bought it, and Harry and I got a very large plywood base on which we set up a simple system. It ran perfectly.

Now I can't remember how it happened, but at that time we got to know an elderly man called Mr. Fischer, who had retired from his job at the Oscar Meyer plant on the north side of Madison. He had an extremely elaborate model railway in his basement. He had switches and sidings and crossovers and bridges, and he taught us all the things we needed to know about model trains. He became very fond of Harry, so we visited him on most Saturday mornings, and he would make extremely generous "swaps" with pieces of equipment. We felt a bit embarrassed by his generosity, but as a small token of our thanks, we always took him a six pack of his favorite beer.

The Lionel train people had introduced something new every Christmas, and Mr. Fischer had it all. He had a circus train where the giraffe automatically bent over when the train entered a tunnel. Also, on command, the train would stop to have its cars loaded with logs and could be made to halt automatically at stations.

Eventually our system also became very complex with many switches and all the other electronically driven devices. Harry got to be quite an expert with electric circuitry, and he has been so ever since.

Helen and Harry were about four and six, respectively, when the circus came to Madison. The performance they liked the best was by the elephants and the elephant trainer. He was tall and handsome and beautifully dressed in silver and gold. It was a bitterly cold day, and when the show was over, we spent a long time trying to get our car started. Eventually we were alone in the car park, cold and dispirited. Seeking help, or perhaps a phone, we walked over to a very large wooden building that was there on the grounds. We opened a door, and there, in all his gold and silver finery, was the elephant man! Further down the room were two elephants. We explained what had happened to us. He was very sympathetic, and while we admired the elephants, he found a can and siphoned it full of gas. Then he walked to our car in the freezing cold, filled our tank, made sure that we were okay, and waved to us as we left. When we got home, the children mentioned the circus, but by far the best, and what they wanted to tell about, was our personal adventure with the elephant man.

Over the years, we had several live-in nannies who looked after Helen and Harry. By far the best was Jean Thain, who had come from England with the recommendation of Harry Fisher's ever-capable neighbor, Mrs. Hester. My daughter, who was five when Jean arrived, is sure that the latter was the incarnation of Mary Poppins. Jean was indeed extremely competent and good with children, and she remained with us for a number of years. Her departure was entirely my doing, although it was for the best of reasons.

My brother Jack had two sons, Michael and Roger. Michael was quite brainy, received a good degree from the university, and he too got a job with ICI. His brother, Roger, had a very inferior job in the same department and became quite depressed. I did what I could to help him, but because I was in the States, I had to do this from a distance. I bought two small reel-to-reel tape recorders, and I sent one to him and we corresponded that way. He told me about his troubles, and I sent him encouraging messages. Roger had a respectable degree in mathematics, so I told him that he did not have to be miserable at ICI working for his brother, that he had a choice in the matter, and that people like him were needed all over the world. Canada, Australia, and New Zealand were possibilities.

It later occurred to me that Roger might get a job in the United States. At that time, we had a large computer center in Madison that my friend Merve Muller directed. He was in need of help, so Merve offered Roger a job. Roger quit his job at ICI, flew over, lived with us, and did well in his new position.

After getting to know one another, Roger and Jean fell in love and eventually married. They moved to Melbourne, Australia, and when we visited them, they were thriving.

Just as my family grew, so did the Statistics Department. By 1968, we had 17 faculty members. Their names and their specialties are listed below:

- George Box—Design of Experiments and Time Series
- John Gurland—Mathematical Statistics and Medical Applications
- Norman Draper—Design of Experiments
- Irwin Guttman—Mathematical Statistics
- George Tiao—Economic Applications
- Sam Wu—Mechanical Engineering
- Bill Hunter—Chemical Engineering
- Donald Watts—Electrical Engineering-Signal Processing
- Jerome Klotz—Mathematical Statistics
- Bernard Harris—Mathematical Statistics
- George Roussas—Design of Experiments
- Richard Johnson—Mathematical Statistics
- Gouri Bhattacharyya—Mathematical Statistics

- Asit Basu—Mathematical Statistics
- Stephen Stigler—Mathematical Statistics and History of Statistics
- Grace Wahba—Mathematical Statistics
- John Van Ryzin—Mathematical Statistics
- Joseph Sedransk—Mathematical Statistics

We also had a number of distinguished visitors, including G. M. Jenkins, R. A. Fisher, G. A. Barnard, D. V. Lindley, J. Durbin, M. Stone, H. Raiffa, R. Schlaiffer, F. Mosteller, J. W. Tukey, F. Anscombe, D. J. Fraser, S. Geisser, and A. Zellner.

Initially Madison had a department that was like no other, with a proper balance between theory and practice. But after a time the department, which had grown to one of the largest statistics departments in the world by the late 1970s, began to change. Ironically I was partially to blame for this. As the head of the new department from 1960 to 1969, I needed to recruit respected senior staff, and at the time, these were from universities such as Berkeley, which emphasized theoretical statistics. Thus, the available recruits were mostly theoreticians who had a totally different view of what statistics was about. What *I* thought statistics was about was solving "real-life" problems in engineering, chemistry, biology, agriculture, etc. At the Porton Experimental Station, and then at ICI, my everyday experience had been employing statistics precisely for this purpose. But a 1978 review of the department conducted by a university committee concluded that "the students feel that applied statistics is important, given the current job market. They believe that some theoretical faculty are not aware of the situation in the outside world and hope that recruiting will maintain the current balance [between applied and mathematical statisticians on the faculty]."[2] Unfortunately, later on the influence of theoreticians outvoted all attempts to keep the department on its original intended path.

Meanwhile, there were other changes to worry about. Nobody had told me that being president of the American Statistical Association was a three-year job. First you are president-elect, then you are president, and

---

[2]Report of the University of Wisconsin Review Committee for the Department of Statistics, June 1, 1978 (the committee consisted of Professors Frank Baker, James Crow, Lawrence Landweber, Morton Rothstein, Howard Thompson, and Hal Winsborough).

then you are past president. All three of these offices have duties attached. There were compensations however. I was president-elect in 1977, when Leslie Kish was president. I greatly enjoyed being his right-hand man, and we became close friends. Leslie had had an interesting career. He fought against Franco in the Spanish Civil War. At one point he was wounded by machine gun fire, but he managed to crawl back to his trench. He was in the hospital for some time, and when he was released, he joined the artillery. That particular war was horrible, and the two sides did awful things to each other. But Leslie was one of the kindest and most considerate men I have ever known, and it was a great pleasure to work with him.[3]

There is a large committee of 30 or so members elected to run the ASA. They meet under the chairmanship of the president at the annual meeting. Somehow in successive years the duration of the meeting had become longer and longer until during my spell as president-elect, it lasted three days. It was easy to see that the main reason that it took so long was that the committee contained a number of self-important people who loved to hear themselves talk. So I prepared myself for being president by reading *Roberts' Rules of Order*, and although I didn't always remember it very well, I spoke confidently as if I did. I found that a loud interjection of "You're out of order!" was often enough to dissuade verbosity. So in the year I was president, I got the meeting over in a day and a half. That year we were meeting in Washington, D.C. Having brought things to a close, I pointed out that the Phillips Gallery was just up the road and that the members should not miss its beautiful exhibits and in particular the gorgeous Renoir, "The Boating Party," at the top of the stairs. I was sure that we would benefit more from this than from further bureaucratic exchange.

At the annual meeting, I gave the president's address, which provided the opportunity to consider the applied versus theoretical camps in our midst. Although many of the theoretical statisticians at Madison and elsewhere were my friends, I had to speak my mind, although humor has a way of softening even a harsh message:

---

[3]Leslie Kish died at 90 on October 7, 2000. See Ivan Fellegi, "Leslie Kish 1910–2000," in *Statisticians in History*, to learn more of Leslie's extraordinary life. This may be viewed on the following website: http://www.amstat.org/about/statisticiansinhistory/index.cfm?fuseaction =biosinfo&BioID=9

When our Executive Director was explaining to me what my presidential duties were, he told me that one of the 'perks' associated with this job is that I get to give the annual address to the Association, and I have a captive audience for as long as they are prepared to sit there. Fred said to me, "George, don't give them anything too technical because this is a light occasion and there will be a lot of people that the statisticians have dragged along—husbands, wives, friends—who have had about all the statistics they can take."

Well, imagine my disappointment. I had prepared a 50-page draft of my talk. It was called "The Present Status of the One-Armed Secretary Problem: A Decision-Theoretic Approach." It was all about sigma fields, Hilbert spaces, and all kinds of squiggly letters with dots on. This I reluctantly set aside. (I don't think any of you would have understood it anyway.) I have had to look for an alternative. I toyed for some time with title, "Whither Statistics?," subtitled "Perhaps We Shouldn't Start from Here," but in the end abandoned that too. Eventually it struck me that many of the issues that we face as members of the American Statistical Association are really not very different from those we face as ordinary human beings. This is what my talk is about.

Some of us have had a preoccupation with optimal or best decision procedures. But the best, of course, is not necessarily very good. For instance, to bring in the aspect of everyday life, if ever I *had* to decide between cutting my throat with a razor blade or with a rusty nail, I suppose I would choose the razor blade. But, although not strictly relevant to the problem as posed, one question that might cross my mind would be, "Have I considered all my options?"

A principle that is being given more attention these days is that of 'robustification.' Here one doesn't attempt to guarantee that things will be optimal over some tractable, but perhaps very narrow, set of circumstances. Instead one tries to ensure that they will be fairly good over a wide range of possibilities *likely to happen in practice*. Look at the human hand, for example. I doubt if there is any single thing that the hand does that could not be done better by some special instrument, but it is very good at doing a very large number of things well that arise in the world as it actually is.

Another way to say this is that there is really nothing wrong with optimization per se, but that we ought to try to optimize over *that distribution of circumstances which the world really presents to us*. The mistake is choosing the best over too narrow a set of alternatives—suboptimization. It is sometimes argued that by doing simplified exercises, we can at least obtain useful pointers. However, I feel that such pointers are very likely to indicate the *wrong* direction, as might be true in the case of the razor blade and the rusty nail.

The difficulty in taking the wider robustification approach is that we cannot expect to get good results unless we are really prepared to engage in the hazardous undertaking of finding out *what the world is really like*. It requires us to have some knowledge of reality.

I believe we do have members, we may have ASA fellows, possible we even had ASA presidents who really do not care what the world is really like. Some years ago, a friend of mine told me about his daughter who was then at oxford University. She was a very bright girl, but she got interested in politics (it was in the 1960s); she got behind in her studies, and the time of graduation was approaching. You may know that in the English system, there are many different grades of bachelor's degrees. The young lady started to worry: Was she going to get a "pass" degree (which is almost like the University spitting at you), or was it to be a third class, lower second class, upper second class, or a first class honours degree? She decided to ask her tutor about it. Finding him buried in the dust of one of the Oxford colleges, she eventually got around to asking him the delicate question, 'Would it matter in the outside world if I didn't get a very good degree?' He looked very startled and said, "*Outside world*? What do *I* know about the outside world?"

It wasn't long after the new department was started that I became afraid that the teaching of the students might be overly theoretical. This prompted me to begin the "Monday Night Beer Session," which met every Monday evening in the basement of my house (Figure 7.3). It was not a formal university course; you got no grades or credits, and you just came when you felt like it. It could be attended by students and faculty from any department.

People came, piled their coats on the ping pong table or on Harry's train set, and we talked. We sat on an odd array of chairs and on an old sofa that had seen better days. There was a cupboard door that we painted black, and it became a substitute blackboard. Svante Wold arrived one evening announcing that he had a surprise for me: It was a chalkboard eraser with a hole drilled in it and a good length of string attached. Not a lot of beer was consumed, but it was always available. Brian Joiner reminds me that I sometimes forgot all about the beer until the last minute, when I would call him and ask him to pick some up on the way over.

GEORGE BOX's

MONDAY NIGHT BEER & STATISTICS

PROFESSOR PETER BOSSCHER

Department of Civil and Environmental Engineering

BAD FOAM BEGATS BAD CARS

Monday, April 4, 1988

7:30 P.M.

2238 Branson Road, Oregon WI

If you need or can give a ride, or need a map, contact Stephen Jones (263-4273).

Everybody is welcome.

FIGURE 7.3
Announcement for Monday Night Beer Session.

Some of the "regulars" at our weekly sessions acted as "talent scouts," looking for people from fields such as engineering, medicine, and business who had problems they wanted to discuss. Usually a graduate student took on this responsibility, and they would visit the engineering school and the College of Agricultural Sciences, as well as other departments, to invite professors to come and speak, or to recommend speakers. When Kevin Little arranged for our speakers, for example, he had Ken Potter from civil engineering speak on time series applications to hydrology, and

Warren Porter came from zoology to talk about modeling problems in animal physiology.[4]

I tried to simulate the experience I had gained in industry by letting students experience the catalysis to discovery that occurred from discussions using statistics. The meetings were a great success. Often people who had brought a problem and received some help would let us know in subsequent sessions how their project was getting on, and a number of discoveries and more than one publication with multiple authors came about as a result. These meetings went on pretty much until my retirement, and former students and other alumni have often told me that what they found most helpful at Madison was the "Monday Night Beer Session."

Much later, my daughter Helen, who gives creative gifts, presented me with a beer-making kit. When we tried it out, things went well until it was time to decant the beer into bottles. Somehow there was an unexpected explosion that left us both covered in beer and reduced to helpless giggles. Later, after I had mastered the bottling stage, I served the homemade brew at the Monday Night Sessions. Unfortunately my students didn't like my homemade brew and I decided that beer making was not my best talent.

In 1966, the General Electric Company offered me $20,000 a year for three years to spend as I saw fit. I used the money to finance the University of Wisconsin Statistical Consulting Laboratory and along with it, the "Statistician in Residence" program. We brought in experienced statisticians whose only duty was to work half time in the lab, and to be available to help anyone at the University with statistical problems. This allowed the students, who spent a semester in the lab as a course requirement, to experience "real-world" statistical problems and to learn from experts to be consultants themselves. From 1967 through 1974, we had six statisticians in residence who came from academia and from industry. These individuals were as follows:

1967–68: J. Stuart Hunter from Princeton University

1968–69: Graham Wilkinson from the Commonwealth Scientific Industrial Research Organization in Adelaide, Australia

---

[4]Personal communication from Kevin Little, May 6, 2012. I am happy to say that Kevin is now reviving the beer and statistics seminar model with students in the Statistics Department.

1969–70: Donald W. Behnken from American Cyanamid Company

1970–71: G. Morris Southward from the Department of Experimental
Statistics at New Mexico State University

1972–73: Harvey Arnold from Oakland University

1973–74: Svante Wold from Umeå University, Sweden

When funding from General Electric ended, the position was supported by the Wisconsin Alumni Research Foundation, which paid half of
the resident statistician's salary. Beginning in 1974, Brian Joiner became
the statistician in residence, and kept the job until 1983. Brian had
spent eight years as a consulting statistician for the National Bureau
of Standards and four years as a professor of statistics and director of
statistical consulting at Pennsylvania State University. He was one of the
three developers of the statistics package Minitab, which he introduced
to the University when he came. Brian had a background in quality-type
statistics, and had first met Dr. W. Edwards Deming in 1963.

As I said earlier, the first course I had to teach at Madison was the
"Advanced Theory of Statistics," later known as course number 709.
Afterward, in the hands of the mathematicians who took it over, this
became a "killer" course, as did its "sequel," course number 710. Both
courses were required of all graduate students in the department, and
both were so difficult that they deterred good students from coming to
Madison. The mathematicians, of course, regarded what I had taught as
a lightweight version of the real thing, when in fact what I was teaching
was a subject that was altogether different from the one they taught. Brian
tried to introduce a new curriculum more in line with the original intent
of the early days of the program, but by then the theoretical statisticians
were in the majority and voted against his proposal. Brian resigned and
formed a successful consulting company that I will speak about shortly.

Despite divisions in the Statistics Department, we came together
for parties and celebrations, the most notable being the festivities that
surrounded the fiftieth anniversary of the department in 2010. Also
memorable were the annual Christmas parties that took place at my
house (Figure 7.3). It became a tradition that the students presented a skit
and the faculty presented a skit. These were high-class performances on
which a good deal of effort was expended. The competition was intense,
and often the students were ahead. This pleased me because I believed

that if you could write a first-class skit, you could also write a first-class thesis. Originality and wit are very close.

An especially good student skit in the late 1970s was based on the first *Star Wars* movie. On another occasion, the faculty skit took place in the years before there were personal computers. The department had a large, stand-alone computer, and a particular student and his professor were known to have a very complicated and special program involving calculations in multidimensional space. It was impressed on all of us that we must on no account do anything that might affect this program.

In the skit, the first scene showed Brian Joiner at the computer at night. He could not resist fiddling with the machine, and suddenly there was a huge flash and all the lights went out. When they were restored, Brian had completely disappeared and it was concluded that he must have gotten lost in n-dimensional space. So in the next scene there was much technical discussion on how to retrieve him. Finally a program was written that would bring him back—one dimension at a time. A very large cardboard box was brought in and connected up. At that time Brian had a beautiful mustache, so after some intensive computing, slowly, from one corner of the box, rose an automobile antenna on the end of which was a lifelike imitation of Brian's mustache. This was the first dimension. After a good deal more discussion and much computing, a sheet of cardboard emerged from one side of the box with a very large photograph of Brian Joiner (Figure 7.4). This was his two-dimensional representation. Finally, after much business and further struggle, there was a flash and the box flew open and Brian jumped out—in all three dimensions!

For Christmas parties, some of us sometimes wrote and performed songs. Norman Draper wrote some very good songs, a memorable one being "The Chairman's Lot is Not a Happy One" based on "A Policeman's Lot is Not a Happy One" from *Pirates of Penzance*. At one party, three former chairmen sang it together. Other songs reflected topical statistical interests. The song "There's No Theorem Like Bayes' Theorem" was written at a time when there was still a great deal of argument about the validity of a particular means of using data to make inferences, sometimes called "inverse probability." It was propounded by a clergyman, the Reverend Thomas Bayes, who lived from 1701 to 1761 and ministered to his flock at Tunbridge Wells, Kent, about 25 miles from

FIGURE 7.4
Brian Joiner retrieved from outer space.

where I was born. He did not publish his ideas, but his friend, Richard Price, communicated them to the Royal Society after Bayes' death.

Since its inception, Bayes's theorem has been looked on with varying degrees of enthusiasm, and other theories—the Neyman-Pearson theory and Fisher's Fiducial Inference—were proposed to avoid the inverse argument. However, a number of statisticians, including myself, came to believe that Bayes had been right to begin with, hence the following, which is sung to the tune of Irving Berlin's "There's No Business Like Show Business":

VERSE (1)

The model, the data you can't wait to see
The theta, beta, sigma, and the rho
The Normal, the Poisson, the Cauchy, the t
The need to specify what you don't know
The likelihood for data you acquire
The perspicacious choosing of the prior

REFRAIN

There's no theorem like Bayes' theorem
Like no theorem we know

Everything about it is appealing
Everything about it is a wow

Let out all that a priori feeling
You've been concealing right up to now

There's no people like Bayes' people
All odd balls from the urn

The other day you thought that you had got it straight
Take my advice and don't celebrate

A paradox by Lyndley could arrive quite late
Another Stone to unturn[5]

REFRAIN

There's no theorem like Bayes' theorem
Like no theorem we know

You can lose forever that perplexed look
If you start to study it right now

Even more enthralling than a sex book
You'll find that textbook
By Box and Tiao

There's no dogma like Bayes' dogma
It's great knowing you're right

---

[5]Lyndley and Stone were statisticians who specialized in Bayesian methods.

We know of a fiducialist who knew the lot
We thought at first he had hit the spot

But after three more seminars we lost the plot
We just could not see the light

REFRAIN

There's no theorem like Bayes' theorem
Like no theorem we know

Fisher felt its use was quite restricted
Except in making family plans for mice

But there, he said, for pinning down a zygote
I'd give it my vote
And not think twice.

There's no answers like Bayes' answers
Transparent, clear and precise

Stein's conundrums you can solve without a blink[6]
Best estimators in half a wink

You can even understand what makes 'em shrink
Their properties are so nice

VERSE (2)

There's Raiffa and Schlaifer, Mosteller and Pratt
There's Geisser, Zellner, Novick, Hill, and Tiao
And these are all people who know what they're at
They represent Statistics' finest flower
And tho' on nothing else could they agree
With us they'd join and sing in harmony.

REFRAIN

There's no theorem like Bayes' theorem
Like no theorem we know

Just recall what Pearson said to Neyman

---

[6]Stein came up with shrinkage estimates redefining Bayesian conclusions using a non-Bayesian argument.

Emerging from a region of type B
"It's difficult explaining to the Lehmann[7]
I fear it lacks Bayes' simplicity."

There's no haters like Bayes' haters
They spit when they see a prior

Be careful when you offer your posterior
They'll try to kick it right through the door

But turn the other cheek if it is not too sore
Of error they may yet tire

REFRAIN
There's no theorem like Bayes' theorem
Like no theorem we know

Critics carp at Bayes' hesitation
Claiming that his doubts on what he'd done

Led to late posthumous publication
We will explain this to everyone
When Bayes got up to Heaven
He asked for an interview

Jehovah quickly told him he had got it right
Bayes popped down earthwards at dead of night

His spectre ceded Richard Price the copyright
It's very strange but it's true!

As I have said, as result of the differences in the department, Brian left the University in 1983 and with his wife, Laurie, started a quality improvement consulting firm called Joiner Associates. The company was tremendously successful, and soon had a world-wide reputation. But by the time the firm was a decade old, Brian became concerned that none of the businesses that he advised had any interest in the pressing environmental issues that were then so much in evidence. In 1996, after attending the World Quality Congress in Japan, he visited China, which he realized was pursuing a resource-devouring industrial strategy that

---

[7]Lehmann was a statistician at Berkeley who did not support Bayes.

mirrored the one in the United States. To clear his head, he and his twin sons David and Kevin went on a 23-day trek through Nepal, where the people seemed content without "stuff." When he returned to the United States., Kevin Little, a statistics department Ph.D. who worked at Joiner Associates, gave him Bill McKibben's book, *Hope, Human and Wild: True Stories of Living Lightly on the Earth*. The book, which describes three communities around the world that addressed environmental problems and required no affluence to develop caring communities, changed his life. Brian and Laurie sold Joiner Associates to the first bidder, and Brian has been a full-time community activist since.[8]

My present feeling is that statistics should be split into two disciplines. One would be theoretical statistics, and the other might perhaps be called "Technometrics."[9] Courses in technometrics would be required not only for chemists, engineers, and so on, but also for mathematical statisticians. Such courses would concentrate on problem solving using the statistical design and analysis of investigations and avoid specious assumptions such as the independence of sequential data.

Some reservations about the manner in which statistics is currently taught are expressed in the following song I composed for a Christmas party. It was based on Gilbert and Sullivan's *I Am the Very Model of a Modern Major General* from the *Pirates of Penzance*:

> *I am the very model of a professor statistical*
> *I understand the theory both exotical and mystical*
>
> *The logic of my argument it is that matters most to me*
> *My chance of making errors is exactly what it's s'posed to be*
>
> *I relentlessly uncover any aberrant contingency*
> *I strangle it with rigor and I stifle it with stringency*
>
> *I understand the different symbols be they Roman, Greek, or Cuneiform*
> *And every distribution from the Cauchy to the uniform*

---

[8]Laurie Joiner (June 6, 1943 — May 21, 2010) had served as the Chief Operating Officer of Joiner Associates. Her premature death after a brief illness was a shock to all of us.
[9]The word was derived partially in response to the title of another journal, *Biometrics*, which deals with biology and statistics. The "techno" in the word refers, of course, to technology.

Chorus

And every distribution from the Cauchy to the uniform
And every distribution from the Cauchy to the uniform
And every distribution from the Cauchy to the uni-uni-form

With derivations rigorous each lemma I can justify
My every estimator I am careful to robustify
In short in matters logical, mathematic, idealistical
I am the very model of a professor statistical

Chorus

In short in matters logical, mathematic, idealistical
He is the very model of a professor statistical

I am the very model of a professor statistical
My art it is immaculate and thoroughly puristical

Judge me by my inner soul and not by my external face
Understand my mind's away in reproducing Hilbert space

With repetitious pleasantries my students all must learn to live
For my wit's one hypothesis to which there's no alternative

I never stoop to folly nor to action reprehensible
I always state assumptions whether ludicrous or sensible

Chorus

He always states assumptions whether ludicrous or sensible
He always states assumptions whether ludicrous or sensible
He always states assumptions whether ludicrous or sensi-sensi-ble

My manner it is modest and not the least hysterical
My errors they are normal both elliptical and spherical

In short in every aspect whether specious or realistical
I am the very model of a professor statistical

Chorus

> *He is the very model of a professor statistical*
> *He is the very model of a professor statistical*
> *He is the very model of a professor statistical*
>
> *I am the very model of a professor statistical*
> *I understand the theory both exotical and mystical*
> *Tho' if all my expounding of the discipline didactical*
> *Could lead to a connection of the theory with the practical*
>
> *If designs that I dub optimal with letters alphabetical*
> *Were readily applicable and not only theoretical*
>
> *If I didn't think consulting was a practice that was better missed*
> *So I could tell a gymnast from a physical geneticist*

Chorus

> *So he could tell a gymnast from a physical geneticist*
> *So he could tell a gymnast from a physical geneticist*
> *So he could tell a gymnast from a physical geneti-neti-cist*
>
> *If decision theory argument could help me make a simple choice*
> *At meetings of the faculty I feel I'd have a stronger voice*
>
> *Why, then, in matters practical, applicable, heuristical*
> *I'd be the very model of a professor statistical*

Chorus

> *Why, then, in matters practical, applicable, heuristical*
> *He'd be the very model of a professor statistical*

Later, I made several trips to countries behind the Iron Curtain, where I saw statisticians struggling with issues that echoed some of our concerns in the department at Madison. In 1982, seven years before the Iron Curtain fell, I was invited to a statistics conference in Bulgaria. The United States would not, of course, allow their citizens to have anything to do with members of the Soviet bloc. But fortunately, I was able to go because I was a British citizen.

Some of the old issues that affected the way statistics was taught in Madison were indeed present in Bulgaria. For example, I had, perhaps, been invited to the conference as a counterblast to the delegates from the Soviet Union. They were all theoretical statisticians and clearly regarded the applied statistics of the Bulgarians with disdain. My hosts particularly liked my questioning the assumptions and general approach of the Russians. I liked the Bulgarians, and when I asked them what life was truly like there, one of them responded simply, "We are all poor but we all have a job."

One reason that the Bulgarians may have had a positive attitude despite the hardships they faced was because they came from a culture that valued humor and story telling. At one point during my visit, we passed through Gabrovo, where men were busy putting up a stage and many rows of chairs. Large signs hung across the road announcing that the following week was the annual National Humor Festival, where people from the audience got onto the stage and competed in telling jokes. There were various categories: children's jokes, mother-in-law jokes, lavatory jokes, and so on, and at the end of the week, prizes were presented to the best joke tellers. There was even a museum dedicated to jokes.

Soon after arriving in Bulgaria, I had been allocated a guide who was very knowledgeable and spoke perfect English. I noticed that when she and I traveled on buses, streetcars, and trains, we did so without paying. When I asked her about this, she said that during the war with Nazi Germany, she had been a member of the Resistance. The Bulgarian government rewarded her with a special pass that allowed her free transport for the rest of her life. I also noticed that whenever we joined a queue in a shop, we were always promoted to the front of the line because the Bulgarians believed this was the polite way to treat foreign visitors.

The Bulgarians were charmingly informal. My guide took me to see an opera in Sofia. Her father was in the orchestra, and when he saw us, he climbed out of the orchestra pit to shake hands and say hello. My guide also had a laissez-faire attitude toward traffic lights. When I exclaimed after we had gone through several red lights, she said, "Well there isn't much traffic on this road."

My hosts arranged that I saw some of the unique features of their country. Our bus went through a long valley in which there were endless plantings of roses, which happened to be at their best, so the air bore a delightful fragrance. I learned that the Bulgarians export attar, an essential oil used in perfume.

We also visited a large castle that was centuries old. The Bulgarians had decided to leave half of it as it was and reconstructed the other half as it had been originally—what a good idea! I'd seen plenty of castle ruins but to see a castle in its pristine glory was an extraordinary experience.

Shortly before I left Bulgaria, one of the professors invited me for dinner with his colleagues in a very modest apartment he shared with his family. The conditions were cramped, with children sleeping in close proximity, but the company was warm. Before leaving that night, I was presented with an unusual gift, a carved head of Bacchus bearing the horns of an animal. In Bulgaria these effigies are hung on grape vines to fend off the attentions of bad spirits. Today Bacchus hangs above the fireplace mantle in my home (Figure 7.5).

A visit to the Soviet Union showed why the Bulgarians relied on humor and good will to survive. To anyone who visited the Soviet Union

FIGURE 7.5
A carved head of Bacchus bearing the horns of an animal above the fireplace mantle in my home.

during its last few decades, its demise was not a surprise. When I was planning to go there for a statistical conference in the 1980s, I was told that the tourist class and first class were pretty bad, but if you traveled "deluxe" it was tolerable. So I went deluxe, and as part of the deal, I was given a choice of a ticket to the opera or the ballet. I had chosen the ballet but was given an opera ticket instead. In the hotel I sought out the In tourist people, who were obliging and friendly and spoke English. There was someone at a desk who dealt with airline tickets and someone else who was in charge of sightseeing and so on. So I found the opera man, and I mentioned that I wanted a ballet ticket instead of an opera ticket and would he please change it. He looked extremely worried and went off to discuss it with one of his colleagues. A lively discussion in Russian went on for several minutes that ended when they all burst out laughing. Eventually my man came over and said, still laughing, "I think the best thing would be if you just went to see the opera. It's very good and you'll enjoy it." I said, "I'm sure that's true, but I would like to go to the ballet. Won't you just change the ticket?" He said, "Well, it's theoretically possible that you *could* change the ticket, but the difficulty is that the opera and the ballet are run by two different Ministries. You would have to make an application to the bureaucracy that controls the opera and then you would need to go to downtown Moscow and wait until the people could see you there and then go up through the system. If you got approval from the head of *that* bureaucracy he would talk to the head of the bureaucracy in charge of the ballet, and if *that* was approved, it would be passed down through the other system and eventually to you and your ticket would be changed. But that would take several weeks to do and we couldn't really be certain of the outcome." We *do* know what the larger outcome was of this bloated and inefficient structure, and we can only conclude that the Soviets were remarkably slow learners.

CHAPTER EIGHT

Time Series

DURING the 1960s I had begun to work on four books that I wrote with five good friends. Norman Draper and I wrote on evolutionary operation, the one book that was published before the decade was over.[1] The second and third books were begun with Gwilym Jenkins and George Tiao, respectively. The fourth, which had the longest gestation period of the four, was *Statistics for Experimenters* with Stu Hunter and Bill Hunter.

When I studied statistics at University College, I remember taking a course on time series. It was all very theoretical, and at the time, I did not see that it had much practical use. Much later, at ICI, I was mostly working on experimental design, but there was a group there from the Intelligence Department that forecasted monthly sales. This involved a panel, one of whose members, for example, knew about the demand in India for indigo, another was an expert on Chinese requirements for certain other dyestuffs, and so on. The forecast was arrived at by putting together the opinions of these individual experts. But when I compared their monthly forecasts with what *actually* happened the following month, I had doubts. The differences between their forecasts and what happened were the forecast errors, and I reasoned that if the forecasts were good, their forecast errors would be unforecastable from past data. I found that for the Intelligence Department's forecasts, this was not true. I went back over the data and found that a simple moving average did better than the "expert" forecast. This discovery did not go well with the members of

---

[1] G E.P. Box and N R. Draper, *Evolutionary Operation: A Statistical Method for Process Improvement*, Wiley, New York, 1969.

An Accidental Statistician: The Life and Memories of George E.P. Box, First Edition. George E.P. Box.
© 2013 John Wiley & Sons, Inc. Published 2013 by John Wiley & Sons, Inc.

the expert panel, but things never erupted into a fight because soon after this, I left ICI to go to Princeton.

Later, the forecasting of time series came up in a quite different context, that of "automatic optimization." Soon after I returned to the United States, one of the problems I was asked about went like this: For a particular process, there is a relationship approximated by a quadratic curve that connected yield y with temperature x. This curve drifted because the catalyst was decaying, and it was impossible to predict which way it would go. So the maximum was not fixed, but it drifted unpredictably. The problem was how to cause the process to change temperature automatically to follow the moving maximum. My idea was to add a sine wave of small amplitude to the temperature x so that instead of remaining constant, it varied sinusoidally about the set point. So if the temperature were not at its optimum, a sine wave was transmitted into the yield. You then looked for this sine wave in the yield y by multiplying y by a second *generated* sine wave (z, say) of the same amplitude and phase, and then you summed the product yz over time. A system was set up so that if the sum Σzy were positive, then the temperature would be automatically increased; if negative, the temperature would be reduced. I have been told that someone else had thought of this idea before me, so no claim of priority is intended.

I thought that helping to design an optimizer of this kind would teach us a good deal, so I tried to get Princeton's Chemical Engineering Department to cooperate in building one, but without success. As it happened, in 1959, Gwilym Jenkins had been visiting the department of statistics at Stanford and had told George Barnard, who had been his thesis advisor, that he was not happy there. George then wrote to me at Princeton, and in particular he said, "Gwilym is very knowledgeable on time series analysis and I would accept his judgment even before John Tukey's." When I showed this to John, he said, "I think we should get this guy here." So that is how Gwilym and I met. Soon after, when I went to Wisconsin, he came to work with me there (Figure 8.1).

At Wisconsin, we met Olaf Hougen, the distinguished chemical engineer who was then nearing retirement. He took to the idea of building the self-optimizing reactor, and he suggested that we immediately apply to the National Science Foundation (NSF) for money to accomplish the task. He added, "I've got two graduate students who can work on it.

FIGURE 8.1
Gwilym Jenkins.

They've both been crossed in love; those are the best kind." (The students were Ken Kotnour and Tony Frey, and they were among a group of chemical engineering students who were later able to use the reactor for their thesis research.)[2]

Our application to the NSF was successful, and we immediately set to work. Olaf Hougen retired, but work on the optimizer continued with

---

[2]K.D. Kotnour, G.E.P. Box, and R.J. Altpeter, "A Discrete Predictor-Controller Applied to Sinusoidal Perturbation Adaptive Optimization," *Instrument Society of America Transactions*, Vol. 5, No. 3, July 1966, pp. 255–262.

the help of Professor Roger Altpeter. I had thought about the automatic optimization problem in a deterministic kind of way, but Gwilym's past experience working on aircraft design showed that it was essential to take into account the dynamic characteristics of the system and of the noise. In particular, the dynamics could cause the transmitted sine wave to be changed in phase and the noise to be "nonstationary."[3] Eventually this taught us much about modeling dynamics and noise with difference equations. We made steady progress, and over a period of about three years, eventually we built the reactor, and got the thing to work.

After a time we realized that our automatic optimization was a particular example of feedback control, and then we saw that the kind of control we were talking about was related to the forecasting of nonstationary time series.[4] So, in this way, Gwilym and I became interested in the general modeling and forecasting of time series.

Previous researchers had emphasized stationary series (i.e., series of observations that varied stably about a fixed mean), but we found that in our research, the stationary model was useless. None of the series that we later encountered from business, industry, or pollution studies, for example, was stationary. We realized that it was nonstationary series that Holt and Winters, and other people in operations research, had tried to forecast empirically, using *exponentially weighted* averages. Their forecasted value was obtained as a weighted average of past values in which the lesser weight was given to more and more remote data. The weights fell off geometrically. This seemed to make sense, but then we realized that this implied a particular nonstationary model described by a simple difference equation. This was the beginning of "ARIMA" time series models. There was, of course, much work on time series that had already been done by Herman Wold and others, particularly for stationary series with autoregressive models, but little on nonstationary models.

I had met Gwilym in the second half of 1959, and our first paper appeared in 1962, so things were moving quite fast.[5] Some considered

---

[3]Nonstationary implied that contrary to widely adopted assumptions, the noise did *not* vary stably about at a fixed origin.

[4]One of the series in the book we later wrote was in fact obtained from the optimizing reactor that we built.

[5]G.E.P. Box and G.M. Jenkins, "Some Statistical Aspects of Optimization and Control," *Journal of the Royal Statistical Society*, Vol. 24, No. 2, 1962, pp. 297–343.

our paper a breakthrough (Johnson and Katz 1992), and our ideas led us to consider a number of other interesting problems. There was, for example, a "golf course problem," named because we used to discuss it as we walked around the local golf course. The problem was to devise an optimal scheme to decide when and by how much you should adjust a nonstationary process that wandered off target. Out of this came what are now called *bounded adjustment charts* for quality control. We assumed a simple nonstationary model, a quadratic loss for the process being off-target, and a fixed cost to adjust it. We solved the problem eventually using dynamic programming, supposing the last observation had just gone over the limit and working backward from there. An important question was how many observations on average would be needed before a process had to be adjusted. We found an approximation that was fairly good, but subsequent research has found better ones. Another problem was the "jam jar" problem, which involved the relation between differential equation models and difference equation models.

Between 1960 and 1970, the work Gwilym and I was doing was funded by the U.S. Air Force Office of Scientific Research (AFOSR). At first we published our results as Air Force reports, but sometime in 1963, Gwilym suggested that we write a book. Initially I was doubtful about this idea because I didn't think that enough people would be interested, but I soon realized that Gwilym was right.

Unfortunately, Gwilym was beginning to have serious problems with what was eventually diagnosed as Hodgkin's disease, at that time, incurable. He went through periods when he was very ill, and then he made temporary recoveries. He was extremely courageous and continued his research and his lecturing until the end of his life, in 1982.

As Gwilym was less able to travel, we still managed to work together. During wintertime, we exchanged tape recordings by airmail. Equations and diagrams were written on pieces of paper that were folded around the tape that we sent. (It was instructive to listen to tapes that we had made earlier and hear ourselves talking about a problem that we had already solved. We always thought: "Why didn't we see that before?")

During the three-month summer breaks, I traveled to England and stayed with Gwilym and his family at their home in Lancaster. Gwilym was a professor at Lancaster University and lived in a beautiful house about 4 miles north of the university. The previous owner of Gwilym's house had kept servants, so I had the maid's quarters on the second floor. I had a room to sleep in and a room to work in. There was a long hallway, and at the other end of the hallway was Gwilym's office. I could go to consult him, and he could come to see me.

Because the research that Gwilym and I did was funded by an AFOSR contract, I got to fly free between England and the United States. on the Military Air Transport Service (MATS). It could be quite an adventure flying on MATS. The service varied considerably in comfort and reliability. You might get a luxurious plane designed for generals—this was much better than flying first class in a civilian plane. On the other hand, you might get a "trooper" that could carry 200 to accommodate the six people who were flying that day. Because of the extremely narrow seats, there was then nowhere to sit comfortably.

At that time I had a little movie camera. One morning, over the mid-Atlantic, there was a rather pretty sunrise. I was filming it when I noticed that one of the four engines of our aircraft had stopped and a propeller was rotating slowly in the wind. When a fellow passenger asked me what I was doing, I said I was photographing the engine that had stopped working. Soon the captain made the announcement that although we could fly perfectly well with three engines, we were going to land at Shannon in Ireland. So we did. The problem at Shannon was that they didn't want us to wander off, so they made us stay for over 12 hours in a large "duty free" area. There, the only popular item for sale was Irish whisky, so some of the passengers were rather high when we finally did get a rescue plane.

Arriving in Lancaster from these journeys was always a welcome experience. Gwilym's house, which stood on its own with large gardens and a fish pond, was surrounded by gorgeous countryside. At first, Gwilym and I used to work in the mornings and go for long walks in the afternoons. After a time, Gwilym was too ill to come with me and I walked on my own. In the valley was the beautiful river Lune, in which I sometimes swam. Most of the time it was deserted except for an occasional salmon fisherman. (In times gone by, the town had been named for the castle on the river Lune—"Lunecaster"—and over the years it had become Lancaster.)

Gwilym was Welsh. He said he didn't speak any English until he was seven years old. His grandmother, who died when she was 102, had not bothered to learn English until she was over 60—she said that hardly anyone spoke English in her village before that. Gwilym had married an English girl, to the disapproval of some of his Welsh relatives. But Meg was a wonderful person and was particularly concerned about Gwilym's health. She believed that the vegetables she grew in her expansive garden would be good for him. Meg's father, Bert Bellingham, another enthusiastic gardener, used to visit periodically to help her, and he and I became great friends.

In the evening, Bert and I would often go to the pub where we'd drink a couple of half-pints of bitter beer accompanied by some crisps—that is, potato chips that were sold in packets for tuppence. We'd been doing this for a year or two during my visits when we discovered that the pub had been taken over by new management that wanted to change it into a "high class establishment." So when Bert went to the bar as usual and ordered, "two halves of bitter and a packet of crisps," the proprietor said, in a somewhat snooty voice, "We don't sell *crisps*." So Bert said, "What *do* you sell?" The proprietor said, "We sell sandwiches." Bert asked, "What *kind* of sandwiches do you sell?" When the proprietor answered, "We sell salmon sandwiches," Bert asked, "How much is a salmon sandwich?" The proprietor then named a price that, compared with our tuppeny crisps, seemed astronomical. So Bert then asked, "What other sandwiches do you have?" And when the proprietor said, "We have lobster sandwiches," Bert immediately asked, "How much are they?" And so it went on. After a series of inquiries of this kind, Bert said, "We don't want any sandwiches," and came back to our table and sat down. When we finished our beer, Bert said, "George, it's your turn to go up." So I went and I said, "We'd like two halves of bitter and a packet of crisps." The proprietor looked at me rather strangely and said, "We don't sell *crisps*." I said, "Oh! What *do* you sell?" and he said, "We have sandwiches." I said, "What kind of sandwiches do you have?" and we went through the list, and I asked him each time how much each sandwich cost, and so forth. When it was all over, I said, "We don't want any sandwiches" and went and sat down again. After that we decided to find a different pub.

Bert had a remarkable ability to make friends. One night when we were going out, Meg asked us to try to get her some tomato plants. At a new-found pub, Bert quickly made himself at home and said, "Does anyone know where we can get some tomato plants?" There was some consultation among those present, and then someone said, "Who you want is old Charlie—comes in around 8:30—he'll be along in about 15 minutes I shouldn't wonder." Sure enough, we went home that evening with some very nice plants at a very reasonable price.

The pub goers often amused themselves with a variety of games. Darts, for example, was a very popular game, and the skill shown by the regulars was impressive. On another of our pub excursions, Bert and I were intrigued by a group of people gathered around the contestants in a game of dominoes. Unexpectedly we saw that the dominoes had nine spots rather than the sort we were used to with six spots. We asked about this, and it turned out that the nine-spot preference existed in certain villages but not in others. This led to lively arguments about the various domino preferences in certain remote hamlets. Whatever the game, our pub keeper often awarded prizes to the victors, which, no doubt, inspired interest as well in the sale of beer.

Bert and I remained good friends long after the summers spent in Lancaster, and whether I visited him at his home, or at the senior housing where he lived later in his life, he always greeted me the same way: "Here's my mate George!"

There was a fish hatchery just down the road from the Jenkins' house, and Meg had acquired about 40 trout there for the fish pond. It was pleasant to watch them swimming around. At that time Meg used to get up early to tend to her youngest child. One morning she said, "Oh Gwilym, I looked out of the window this morning and there was this huge bird catching and eating our fish." Well, Meg was a country girl and she went into the village the next day and came back with a rifle she had borrowed. The farmer who had leant it to her told her that the bird she had seen was a heron and that she should wait until the bird came again and then fire the gun in the air to frighten it away. Gwilym was horrified. He said, "Oh but Meg, we can't have a gun in the house." And they had a disagreement. So I said, "I've been in the Army and if you like, I'll fire the gun." So that was agreed upon, and a few mornings later Meg woke me up at about 6 a.m. and said, "Oh George it's here." I looked out of the window, and sure enough, there it was quietly standing very tall on the edge of the pond catching our fish. So I fired the gun into the air and the heron took off, and so far as I know, it never came back.

This got Meg thinking that some trout might make a very good dinner, so we set about catching some. This quickly became a joke. So far as I can remember, Gwilym had a worm on a bent pin on the end of a piece of string. Meg had a net, and I had a rectangular piece of wire netting. My plan was to wait until a fish got into a corner and then put my netting down so that it couldn't get out. We all had a great time trying, but as we quickly found out, none of this was any good: The fish were much too fast for us. The problem was solved when Meg discovered that the village postman was an avid angler. He quickly caught us some fish, and we had some delicious dinners.

A later problem that Gwilym and I worked on concerned the forecasting of *seasonal* time series. At that time, I was doing some work for Arthur D. Little, the well-known consulting firm, and they were forecasting using exponential discounting with polynomial models, but they were not producing very good results. For practice we studied some data of R. G. Brown, which showed for each month over a period of several years the number of passengers traveling on transatlantic airlines. This series had a 12-month developing pattern that was low in the winter and high in the summer and varied somewhat every year. This led us to devise what came to be called the "airline model." We got a big bang for our buck from this kind of model because even though it contained only two parameters (unknown constants to be estimated from past data), it produced a seasonal pattern made up of 12 sine waves of different frequencies. These tracked the data very well and could develop and change as new data became available.

Gwilym and I had both learned a lot about likelihood from George Barnard, and it seemed natural to use likelihood to estimate the coefficients. But at that time, I was getting more and more interested in estimation using Bayes' theorem (I thought of likelihood as being a timid man's way of doing Bayes), but for samples of the size we needed, it didn't make much difference anyway.

A problem of estimation with any kind of time series is that what happens at the beginning depends on what had happened before you started! This is, of course, unknown. However, we realized our models were reversible, and so a natural thing was to forecast what you didn't know by starting at the far end of the series, and "back forecasting" the whole series, including the few preliminary values you needed to get started. We tested this idea on a number of series and it was later validated from a mathematical point of view.

Sometimes you needed to model the interrelationships between several related time series. For example a famous series shows the supply of hogs, the price of hogs, the price of corn, the supply of corn, and the amount of farm wages over a number of decades.[6] Taking into account the dependencies of these five series produced much more accurate forecasts of each one of them. These ideas were widely applicable, and as a result, George Tiao and I got some students programming multiple time series estimation at Madison. This work was further developed in Chicago under the direction of Professor Lon-Mu Liu. In addition, Gwilym developed a program for multiple time series independently at the University of Lancaster.

At the University, Gwilym had started the Department of Systems Engineering, and he and his students had developed close associations with local industry. The students solved problems with scientists working at these industries as part of their degree requirements. Gwilym's department was very successful and brought in a good deal of money to the University. But people in other departments became jealous, and there were many difficulties. So later, in 1974, Gwilym left the University and set up his own company (Gwilym Jenkins and Partners) where his ideas were further developed.

Gwilym chose a publisher, Holden-Day, for our time series book. They were responsible for the first edition of *Time Series Analysis: Forecasting and Control* that appeared in 1970. Unfortunately, Holden-Day was quite reluctant to pay royalties.

One day Gwilym asked me to call his lawyer in San Francisco about this. I asked him how he had found the lawyer, whose name was Norman Macleod, and he said that he had reached him through the British Consulate. So I called up this lawyer. He had an extremely strong British accent and said, "Oh, Professor Box, I'm terribly glad you called me. Gwilym said you would." I said he didn't sound to me like an American lawyer, and he said, "Of course I used to be in England, but I took the Bar Exams and I am working over here now." He cleverly arranged for a

---

[6]U.S. Department of Commerce, *Statistical Abstract of the United States, 1950*. This is a five-dimensional multiple time series consisting of 82 annual observations from 1867 to 1948 of factors pertaining to the hog trade in the United States.

permanent pending legal action that we could threaten to use whenever we didn't get paid. Later Holden-Day went bankrupt.

Sadly much of the strife with Holden-Day occurred while Gwilym's health was rapidly declining. While before we had sent tapes on which we discussed our ideas while writing the book, as the illness progressed, I added tapes of the old Goon Show as well as tapes containing jazz and other music that Gwilym loved. I recently found a copy of a letter that I sent to him in the spring of 1982, shortly before he died. In it I wrote:

> Last week I was teaching a short course on time series for some Army Operation Research officers. . . . I always do this kind of thing conditional on everyone having a copy of our book and when they all brought them up to be autographed I told them a bit about you and what fun we had together writing it. When we did the transfer function example someone asked why we read the records at nine second intervals and I told them that, so far as I could remember, it was because Meg, who was helping us that day, said that that was the shortest interval that she could read and we said: 'Then that's the one we want.'
>
> I often think of you Gwilym while I am walking Victor, our dog, up near the golf course. I remember how we would go for walks on the golf course and try to sort out 'the jam jar model.' And how I got very fed up one day and you saw a hill on the opposite side of the lake and you said, 'Let's go over there on the other side of the lake and climb up that hill,' and we did. I remember too all the walks we took from Halton Green House and I can see in my mind exactly what the house looks like--from below as you're walking on the down-below-road, and from above, when you come over from the farm, and what the driveway looks like and how we would sometimes see Meg working away in the garden. The trout pond—I remember the time we tried three different methods for catching them without success. I was sure that a sieve on the end of a pole would work but the trout had other ideas. The ducks came later—I think after the heron that we frightened with a shot gun.
>
> Very soon now I will be setting off for Bulgaria and returning via England so I do hope that I can come and see you . . .
>
> With Much Love,
>
> GEORGE

The book appeared in subsequent editions, published first by Prentice-Hall and then by John Wiley with Gregory Reinsel, my colleague at

Madison, as a co-author.[7] It is now in its fourth edition and has been translated into many languages. When it was first published, the reaction of reviewers was negative. Some people said it wasn't rigorous enough; others said there was nothing new in it. However, we were not overly concerned because we knew[8] that initially, original work was invariably met with hostility. For example, the first paper on response surfaces, and the paper in which the word "robustness" appeared for the first time, were both quite difficult to get published. I think new ideas upset people. For the time series book, there was much discussion, for example, about our use of differencing. All we were saying, really, was that you could be better off modeling the *rates* that things happened. Now, however, as a direct consequence of these ideas, cointegration and unit roots are a big business in econometrics.

On the first page of the book, we mention five explicit applications:

1. The forecasting of future values of a time series.
2. The determination of how the output is dynamically related to the input for a system subject to inertia.
3. Determining the effect of intervening events on the behavior of a time series.
4. The representation of relationships among several time series.
5. The design of control schemes for compensating deviations from a desired target.

Thus, what came out of the research on the automatic optimizer was much more than was our initial intention. This confirmed what we believed: That the best way to develop theory is to study practical examples carefully.

In addition to its impact on our ideas about control, the book has had a large influence on economics and business. Also, econometric models have a lot in common with chemical kinetic models used in describing complex reaction systems. Gwilym and I both had some experience with these. Later, several quite different theoretical kinetic models were put

---

[7]Tragically we lost Greg on May 5, 2004 at the young age of 56. He was a highly respected member of the Statistics Department for 28 years.
[8]And were told earlier by R. A. Fisher.

forward by different sets of chemical engineers and chemists to explain the production of ozone and the subsequent pollution of the atmosphere. The problem was that all the kinetic models contained very large numbers of parameters, which it was quite impossible to estimate from the available data—the same problem that arose in economic modeling. We believed that a good approach is that, rather than start with the model, you should start from the *data* and produce a simple dynamic-stochastic model empirically, and *then* try to relate this to theoretical mechanisms that could be identified. This has turned out to be a very valuable approach. At one time we had a joint project along these lines with the economists at Madison, but I am sorry to say that nothing came of it.

When it came time for my students to pick a thesis topic, I usually discussed their interests with them and then suggested possibilities for further pursuit. Three of my students, however, came to Madison with a predetermined interest in time series. Earlier, Dean Wichern, Paul Newbold, and Larry Haugh had written good theses on the subject. As a matter of timing, however, the students who came after *Time Series* was published in 1970 had more familiarity with the subject. These included Bovas Abraham, Johannes Ledolter, and Greta Ljung.

Bovas Abraham was from southern India. Before coming to Madison in 1971, he had received a Master's degree in statistics from the University of Kerala, where, for a time, he also taught, and he had also spent two years teaching secondary school in Cape Coast, Ghana. He had then received another master's degree from the University of Guelph. Bovas had received a partial research assistantship from Wisconsin, and shortly after moving to Madison, he came to talk to me about it. Before he left my office, he agreed to plot some difficult diagrams for me. Only much later did he tell me that he had had no idea how to do this. He persisted, however, and with great enterprise, he eventually discovered a computer program that could plot the graphs. After a year, Bovas passed his qualifying exams and asked me if I would supervise his dissertation.[9]

I insisted that my students produce a well-written thesis, and worked hard with both native and non-native speakers of English to make them understand the importance of this. Because I traveled a great deal in the

---

[9]B. Abraham, *Linear Models, Time Series and Outliers*, Ph.D. dissertation, University of Wisconsin, Madison 1975.

1970s, I often read parts of a student's thesis and recorded my reactions, as well as suggestions for improving the writing, on a tape that I would then give to the student. This worked surprisingly well. Bovas and I used this technique, and we also had many meetings. When I was traveling during the week, I often met with students on the weekend. Several students would car pool and make the drive to my home south of Madison. Bovas wrote a good thesis and was able to publish a number of papers that were based at least in part on this research.[10] Later, he became professor of statistics at the University of Waterloo in Ontario.

In 2000, Bovas nominated me to receive an honorary doctorate in mathematics at his university. We had a wonderful party at his home on the evening of the convocation. Bovas' wife, Annamma, who is a first-class cook of, Indian food, was dressed traditionally and looked exquisite. While we were there, they were planning the wedding of one of their two daughters. Annamma had been to the hotel where the reception was to take place to instruct the kitchen staff on Indian cooking and in particular, what rice to use and just how to cook it. When we left Waterloo, she sent us home with a jar of very spicy Indian pickles and an assortment of all the right spices for curry.

While he was a student in Madison, Bovas became friends with another of my students, Johannes Ledolter. Johannes, or Hannes, as he was often called, had come from Austria, and he had planned to stay in Madison no longer than a year. He came to my office, a fresh-faced 21-year-old, and asked me about the time series course I was about to teach that semester. I lent him a copy of the book that Gwylim and I had published saying, "Take a look at this. If you like the book, you'll like the course." He liked the book, took the course, liked that as well, and decided to stay as my graduate student. He became my teaching assistant for the time series course two years in a row, and he joked that he had now taken the class three times. Not surprisingly, he also wrote his thesis on time series, but when he was partway through the process, the Austrian Army called him up for mandatory service. Over the years, I had written a

---

[10]See, for example, B. Abraham and G.E.P. Box, "Linear Models and Spurious Observations," Applied Statistics, Vol. 27, No. 2, 1978, pp. 131–138; and B Abraham and G E.P. Box, "Bayesian Analysis of Some Outlier Problems in Time Series," *Biometrika*, Vol. 66, No. 2, 1979, pp. 229–236.

number of letters to help students with various visa problems, but this was the first time, I had had to ask an army to forgive someone their service.

Hannes, of course, shared with Bovas an interest in time series. After Bovas went to teach in Canada, and Hannes became a professor at the University of Iowa, the two worked together for many years, a friendship and professional partnership that started when the paths of a man from Southern India and a near youth from Austria crossed paths in Madison in the fall of 1971.[11]

At about this time, Greta Ljung came to Madison to study statistics and was another Ph.D. student interested in time series. Greta had come from Finland, and while many people from Finland learn English from a young age, hers was exceptional. Moreover, of all the students I ever had, she excelled as a writer. When it came time for her to write her thesis, it was smooth sailing. She and I published a number of papers, one of which included a write-up of the so-called Ljung–Box test.[12]

Greta taught for a number of years, at MIT, but in recent years, she has been a principal scientist at AIR Worldwide Corporation, which leads the catastrophe modeling industry. There she heads a team that does statistical modeling of tropical cyclones, severe thunderstorms, and wildfires.

---

[11] B. Abraham and J. Ledolter, *Statistical Methods for Forecasting*, 2nd ed., Wiley-Interscience, New York, 2005; and B. Abraham and J. Ledolter, *Introduction to Regression Modeling*, Thomsen Brooks/ Cole, Belmont, CA, 2006.

[12] G.M. Ljung and G.E.P. Box "On a Measure of a Lack of Fit in Time Series Models," *Biometrika*, Vol. 65, No. 2, 1978, pp. 297–303.

CHAPTER NINE

George Tiao and the Bayes Book

GEORGE TIAO received his MBA from New York University in 1958, and then he came to Wisconsin with the intention of getting a Ph.D. in international finance. Eventually he pursued econometrics and was one of the students in the first course I taught. He was very interested in Bayes, and while he was a student, he and I wrote an article about Bayesian methods for *Biometrika*.[1] His thesis was on the robustness of linear models from a Bayesian viewpoint.[2]

When Bayes, your conclusions need not be based on the assumptions that the errors were normally and identically distributed and independent. You could for example explore:

1. What effect would various types of departures from the normal law have on the standard tests? This would include the effect of a lack of independence between the errors.

2. How could efficient procedures be devised appropriate for such circumstances?

It was clear from the beginning that George would make an important contribution to our growing department, so when he completed the Ph.D. in 1962, I offered him an assistant professorship with a joint appointment with the Business School. In 1965–1966, when I was invited to spend a year at the Harvard Business School, I arranged for him to come too,

---

[1] G.E.P. Box and G.C. Tiao, 'A Further Look at Robustness Via Bayes' Theorem. *Biometrika*, Vol. 49, 1961, pp. 419–432.

[2] G. C. Tiao, Bayesian Assessment of Statistical Assumptions, Ph.D. Economics, University of Wisconsin, 1962.

An Accidental Statistician: The Life and Memories of George E.P. Box, First Edition. George E.P. Box.
© 2013 John Wiley & Sons, Inc. Published 2013 by John Wiley & Sons, Inc.

which provided an opportunity for us to work together on a book about Bayesian inference.[3]

George searched housing for our respective families, and he found two places that would do. One was the old, dark house with many rooms that I mentioned earlier, where we lived. George and his family stayed in a more reasonably priced home that was not far away. We set up an office in one of the bedrooms where we could lay out our papers, books, and scribbled notes. By the end of the year, the book was well on its way.

George Tiao's logical mind is illustrated in the following story. While in Cambridge, we discovered that the Harvard campus has strict rules about parking. Once, George and I had to bring a large and heavy tape recorder from our car to the business school, so we tried to park as close as we could. We found the perfect spot, but it was illegal to park there. A sign warned that the first-time offender would receive an orange ticket, the second-time offender would get a red ticket, and a third offense revoked all parking privileges *any where* on the Harvard campus. When I told George that we couldn't park there, he said, "Yes we can—twice."

We returned to Madison with plans to work on the book whenever possible, but the pace of academic life made it difficult to have sustained periods of collaboration. In 1968, however, George and I wrote a paper on Bayes and outliers that generated a good deal of interest.[4] Finally, we got our chance to finish the book in 1970–1971, while we were both on leave at the University of Essex.

In 1963, a new university was started at Essex, in southern England, and in 1966 George Barnard left Imperial College to become chair of the mathematics department there. In 1970–1971, he invited me to come to Essex for the academic year. At the time I had three Ph.D. students (Larry Haugh, Hiro Kanemasu, and John MacGregor) and I was partway through the Bayes research with George. George Tiao decided to go to Essex as well, and he had two Ph.D. students (William Cleveland and David Pack). George warmly welcomed the large entourage that went in the end: George Tiao and I went with our respective families, the five graduate students, and some of *their* families as well. Bill Cleveland's wife gave birth to their first child there, and Larry Haugh went with his

[3] *Bayesian Inference in Statistical Analysis*, John Wiley and Sons, New York, 1973.
[4] G.E.P. Box and G.C. Tiao, "A Bayesian approach to Some Outlier Problems," *Biometrika*, Vol. 55 (1), 1968, pp. 119–129.

wife, Jane, and one-year-old daughter. Larry recalls that the Americans in our group sometimes had to "rough it" in England: "Jane and I lived in a small village, Wivenhoe, which was just an English country path (across farm fields) … away from the Essex campus. … Living in the English village was quite the experience, … wearing sweaters so the thermostat could be lowered, living thru [sic] the postal strike and various rail strikes."[5] None of us knew that six years later, Jimmy Carter would be urging Americans to keep *their* thermostats down.

When we arrived in Essex, at George Barnard's suggestion, we met each week for a seminar studying Fisher's earliest papers. Each of us studied one of the articles and then presented it to the others. When Fisher first went to Rothamsted in 1919, he had been asked to review the data from the long-term Broadbalk soil fertility experiment, which had begun in the mid-nineteenth century and continues to the present day. A large area of land planted in wheat had been divided into long strips, each of which had been treated with the same nutrients (N, P, K, etc.) since 1843. Differences in the resulting wheat crops were plain to see, but it had been Fisher's task to determine whether further information could be gleaned from such experiments. Fisher's response to this and many other problems can be found in a series of papers called "Studies in Crop Variation."

What we learned from these papers exceeded all expectations. There were many things we had not associated with Fisher and many that we believed were of a much later date. For example, there was an *analysis of residuals* (including how they are autocorrelated), the use of *distributed lag models*, and the *multivariate distribution* of *regression coefficients*. Fisher later referred to these studies as "raking over the muck heap," for these investigations made him realize that often desired information was just not available from data such as was supplied by the Broadbalk trials. This led to his inventing the design of experiments[6] that made it possible to study efficiently and simultaneously those specific effects that *were of interest to the experimenter*. The design of experiments later became a research area of great interest to me and my students.

---

[5]Personal communication from Larry Haugh, March 30, 2012.
[6]R.A. Fisher, "The Arrangement of Field Experiments," *The Journal of the Ministry of Agriculture*, Vol. 33, 1926, pp. 503–513 was the first article to discuss the design of experiments.

Winter weather in England can be cold and dispiriting so it was not a surprise when two of the students, Hiro Kanemasu and David Pack asked if there was anywhere they might go over Christmas where it was *warm*. I suggested Spain, and they went there, but before they were due to come back, they ran out of money. We sent funds via American Express, but because of some complicated snafu, the money didn't arrive. We sent some more, but they never received that either. They later told us that they got so hungry toward the end that they got into trouble sitting in restaurants and eating the sugar.

Hiro had not finished his thesis when he went to work for the World Bank, so periodically he sent me new sections handwritten on yellow sheets from Washington. I got one batch that was covered with tire marks extending in all directions over the pages. He tended to be forgetful, and one day he had left a large chunk of his thesis on top of his car. The pages had blown all over Pennsylvania Avenue, and Hiro had spent a large part of the morning running in and out of the traffic recovering them page by page. Although at that stage his thesis was hard to read, it was good stuff. Titled *Topics in Model Building*, it was successfully completed in 1973 and Hiro received his Ph.D.

Smog had plagued Los Angeles since the early 1940s. Scientists had been measuring the concentrations of a number of pollutants for many years, so they had a tremendous amount of data. Beginning in 1973, George Tiao and I worked closely with them and particularly with an excellent chemist, Walter J. Hamming. By the time we met him, Hamming was the Chief Air Pollution Analyst for the Los Angeles County Air Pollution Control District. To deal with pollution, the state of California had introduced various laws, for example, outlawing open burning in incinerators and controlling certain industrial emissions. Our friend Hamming, however, was certain that almost all the pollution was what came out of the back end of an automobile, although at that time nobody wanted to believe it.

In 1966, California adopted the nation's first emission standards for hydrocarbons and carbon monoxide. George and I developed intervention analysis initially to look for a significant change in the pollutants at the time of the enactment of this law, and we found it. It became indisputably clear that Hamming was right.

The theoretical aspects of this kind of "intervention analysis" were discussed in a paper that George Tiao and I had published.[7] This concerned the general problem of estimating a change in level that might have occurred at a known point in a nonstationary time series. For the Los Angeles data the change in level may be estimated by a linear combination of the data for which the weight function consisted of two exponentials back to back. The first had positive weights, and the second, had negative weights. This was very sensible because what happened just before and just after the event got the largest weights and were obviously of most importance.[8]

George taught at Madison from 1962 to 1982, when he went to the University of Chicago. Subsequently he has spent a great deal of time working in Taiwan, where he is held in very high regard. Through the years, our friendship has remained strong. When I turned 80 in 1999, George organized and hosted a wonderful birthday party at the Fortnightly restaurant in Chicago. Old friends and colleagues came from afar. It seemed only appropriate to sing them "There's No Theorem Like Bayes' Theorem."

---

[7]G.E.P. Box and G.C. Tiao, "Intervention Analysis with Applications to Economic and Environmental Problems," *Journal of the American Statistical Association*, Vol. 70, No. 349, 1975, pp. 70–79.

[8]G.C. Tiao, G.E.P. Box, and W.J. Hamming, "A Statistical Analysis of the Los Angeles Ambient Carbon Monoxide Data 1955–1972," *Journal of the Air Pollution Control Association*, Vol. 25, No. 11, Nov. 1975, pp. 1129–1136.

"There are 364 days when you might get unbirthday presents, and only 1 for birthday presents. you know."

CHAPTER TEN

Growing Up (Helen and Harry)

$\mathcal{H}$ELEN and Harry had an especially memorable year at Essex. There we were offered rooms at Stansted Hall, about 12 miles from the college. This was a castle that dated from the fourteenth century and belonged to the Earls of Essex. It still had a moat, a long driveway and gate house, and extensive gardens. Our rooms were on the third floor—formerly the servant's quarters—and you had to be careful not to hurt your head on the enormous beams that held up the ceiling. The castle now belonged to Lord Butler, who had been a recent candidate for Prime Minister. He gave it to the University to be used as a residence for visiting professors and their families.

At that time there was a special deal whereby you could buy a car in England, drive it for a year, and then take it back to America duty free. On this scheme, I purchased a sturdy vehicle, a left-hand drive Volvo station wagon. After it arrived, I used it to navigate every day, sitting on the wrong side of the car over the 12 miles of hilly, narrow, and bending road from the University to Stansted Hall. We brought the car back to Madison where it performed admirably in our frigid winters but cost a small fortune to repair.

A Professor Hill from Canada was also visiting the University, and he and his family lived in a different part of the castle. They had three children, Caroline, Simon, and Benjamin, who were close in age to Helen and Harry, who were then nine and seven. We all thought it was very romantic to live in a castle, and since I had a movie camera, we decided to make a film called "The Blue Knight." Young Caroline played the

An Accidental Statistician: The Life and Memories of George E.P. Box*, First Edition. George E.P. Box.
© 2013 John Wiley & Sons, Inc. Published 2013 by John Wiley & Sons, Inc.

princess, Helen was the wicked witch, Harry was the prince or "blue knight," and I was the cameraman.

The movie opens with the handsome prince taking a rose to his love at a window in the castle. The pair have to meet secretly because the princess has a cruel uncle who guarded her closely. Their idyll is cut short when the prince is called away to fight in battle. As the princess walks back to the castle alone, the witch and her assistant, played by Caroline's brother Benjamin, kidnap her. When the prince finally returns to his love's window, he realizes that she is gone. At this point, the film cuts back to the witch who is throwing the princess into a dungeon (a real one). The noble prince quickly rushes to rescue her, but on the way, he traverses a dense forest where he encounters a terrifying dragon (played by Mrs. Hill in the garden in a green mackintosh and red ski mask), whom he must fight to the death. During the battle, he loses his sword and there is a brilliant camera shot that shows shadows of the duel falling on the sword. The prince recovers his sword and slays the dragon, but no sooner does he accomplish this than he encounters an enormous bear played by Simon Hill wearing a fur coat and hat. Unfortunately the bear's hat falls off during this scene, but the filming continues nonetheless. The prince dispatches the bear and continues his travels, only to arrive at the wrong dungeon. He calls out his lover's name, but the witch hears him and, arriving with a ring of enormous keys, which looked to be original with the castle, locks him into the dungeon. With great skill, however, the prince is able to sneak out when the witch brings him bread and water, but she absentmindedly leaves the keys in the lock (at this juncture, the prince somewhat overacts, using greatly exaggerated motions to effect his escape). The film then cuts to the prince and princess walking in the garden, making it clear that the rescue has been successful. But suddenly the witch emerges from a well, and another battle ensues during which the prince manages to push the witch down the well, and she is never heard from again. Unfortunately the cruel uncle continues to forbid the love between prince and princess. All is saved, however, when a fairy (also played by Helen) appears and announces that because the prince has rid the kingdom of the witch, they can finally marry.

As a child I had a pretty good imagination, and I hoped that Helen and Harry would have the same. They have told me that I was a very good story teller at bed time. I also read books to them, especially *Alice in*

Wonderland, which I have read many times since my grandmother first read it to me.

Alice is the perfect young woman, brave and independent. Moreover, there are phrases in the book that all of us, and perhaps scientists above all, should keep in mind. These include:

If you don't know where you are going, any road will get you there.
His answer trickled through my head like water through a sieve.
It's a poor sort of memory that only works backwards.

When my daughter Helen was about 14, she and her mother, Joan, became increasingly unhappy with one another and Helen decided she wanted to go away to a boarding school. Joan and I visited a number of schools within a 250–mile radius of Madison, and we finally agreed that by far the best was Culver Military Academy in Indiana. This was a boys' school, but they had recently started a program for girls. We were impressed with the small classes, the high academic standards, and in particular, the English department, which happened to be run by a man from my hometown in England.

The boys had four companies: the infantry, the artillery, the cavalry, and the band. The cavalry was the "black horse troop" with an assembly of beautiful black horses.[1] The artillery were motorized and had small field guns. Although the boys' program was firmly established, they didn't seem to know what to do with the girls. Helen quickly acquired a brilliant academic record, but she despised discipline. For example, she liked to take long walks, and although the grounds at Culver were very extensive and clearly marked, she was continually being found in places she should not have been. I think her attitude puzzled the people at Culver, and it seemed that whenever we went to see her, she was being punished with "KP" duty, peeling potatoes.

Harry came with us on our visits to see Helen. Every Sunday the boys put on a spectacular parade, and Harry was very taken with this. I say "the boys" because there was not a single adult involved. The parade was impressive, with all four companies beautifully coordinated: The band played, the infantry marched, and the black horse troop and artillery made complicated maneuvers, all under the direction of the senior boys.

---

[1] The Black Horse Troop has taken part in nearly every presidential parade in Washington, D.C. since the 1913 inauguration of Woodrow Wilson.

Seeing all of this, Harry decided that he wanted to go to Culver too. I tried to tell him what he would be getting into. The boys' stuff was serious. For example, a first-year student (a plebe) was a nobody who had to call everyone who was his senior "Sir." Also he had to march, rather than walk, everywhere he went. This was whether he was inside the school or outside, and he had to make smart military turns with stomping feet as he rounded corners. All the military requirements, I explained, were in addition to academic duties. But Harry said he understood and wanted to go anyway.

The superintendent at Culver was an ex–colonel. He appeared at the Sunday church parade impeccably dressed in a snow-white uniform. Acknowledging the parents entering the church, he stood in the transept and close by was a special memorial brass plate built into the floor on which no one was allowed to stand. After we had decided that he could go to Culver, Harry came with us one Sunday to visit Helen. That day he looked unusually scruffy, but he went straight to the colonel, and standing on the forbidden spot, gave him the good news that he was coming to Culver next year. The colonel seemed to clear his throat several times, but I didn't quite catch his words.

Harry loved everything about Culver, and although he was in the infantry rather than the band, he played bass in a jazz band that was in considerable demand. When he graduated, he was a lieutenant, next to the highest rank. (There was one captain and one other lieutenant.) Helen graduated at the top of her class academically, but I think close to the bottom in matters involving discipline.

During the summer, Harry got himself a job in Culver's flying program, run by a very acerbic ex-Air Force colonel who had flown B17s. Harry's duties were to gas up the small planes used for training and to operate the radio. He reported to the colonel, who emphasized that on no account was he to be disturbed until he had had his morning coffee, and that the staff should never do anything to cause adverse publicity for the flying school.

The students under instruction had flying lessons quite early in the morning, and one day Harry received a radio message that because of some oversight, one of the planes had come down in a field. So following instructions, Harry said nothing about this until he had brought the colonel his coffee. When Harry finally did tell him what had happened, he blew a fuse. "What!" he said. "Oh my God. Get into the car and drive me there." They found the plane, which, along with its occupants, was unharmed. The colonel had them push the plane across the field and onto the road, and he then climbed into the cockpit and took off using the road as his runway. Fortunately, there was not much traffic and the plane landed safely at the flying school. More important, however, was the lesson Harry learned about what might take precedence over morning coffee. Harry later became an expert pilot and was himself a flight instructor.

I spent many years as a spare-time consultant for various companies, and this, of course, provided income on top of my professor's salary. I saved these funds for the children, and eventually I sought legal advice as to how I might set up a trust for them. I had the assistance of a very competent lawyer, Ralph Axley, who urged me to word the legal documents in such a way that the children would not be restricted in how the funds could help them. If one wanted to go to college, the money was there; if another wanted to be an artist or a carpenter, the money was there too. In the end, both children went to university and received advanced degrees.

Helen went first to Oberlin and then to the University of Wisconsin Medical School. After surviving her residency in Eau Claire, she and some of the other new doctors made a film depicting their harried and sleepless existence as residents. They somehow managed to shoot this in the hospital, and in one scene, they are featured wheeling around from patient to patient on roller skates. In another, Helen is catching forty winks in a cot when the phone rings. "Take two aspirin and go back to sleep," she mutters groggily. Today Helen is a physician in Chicago. She works in a clinic that serves a largely Latino population that, without the clinic, would have little access to health care.

Helen's husband, Tom Murtha, has been a city planner for most of his professional life and currently works in Chicago on the challenging issues related to traffic congestion. The development of bike trails is a special interest of Tom's, and he often rides to work from Oak Park to the downtown.

Helen and Tom have two bright and energetic boys, Isaac Alexander and Andrew Jefferson. Isaac, 16, loves theater, musicals, and literature and has an avid interest in politics. Last summer we attended a production in Madison, *The Lamentable Tragedy of Scott Walker*, by Doug Reed. Isaac understood all the jokes and the more subtle satire and suggested to his drama teacher that they perform the play in Oak Park. In a recent phone conversation, he reminded me, "Don't forget to give Bert [our cat] credit in your book!"

At 14, Andy has a lot of energy and is very interested in sports. After four-and-a-half years of Aikido, he is now in adult classes, and he plays soccer, volleyball, ping-pong, and all sorts of pick-up games with friends. He manages a busy schedule that makes me tired just thinking about it. He is on to high school this fall and, at this point, wants to be a neurosurgeon. If he continues to want to do this, I have no doubt that he will.

Harry went to film school at the University of Texas in Austin and graduated with a Master's degree. In 1993, having become especially interested in lighting, he wrote an encyclopedic book on the subject that has gone through four editions.[2] He also learned to be a pilot with his portion of the money.

Some years ago when we were in California, we visited Harry on the set of the Disney television series, *Even Stevens*. Harry was shooting an episode, and we spent the entire day watching. We were impressed with what a slow process it was, with so many takes, and the very large number of people who seemed to spend a lot of time just standing around.

While we were there, they were shooting a scene that took place at a school where one of the students had brought her pet pig to show the class. She was sitting in the playground with a boy who was eating his lunch, and the pig was tethered nearby. Harry's camera was on tracks, and it moved forward to show the boy and the girl deep in a conversation that so absorbed them that neither of them noticed that the pig was eating the boy's lunch. The success of the scene depended on the pig's appetite. Earlier mistakes must have been made because the second assistant director went round shouting through her megaphone, "Please! Don't anybody feed the pig!"

Harry met his wife, Stacey Kosier, in Hollywood, and together they bought and fixed up a bungalow not far from Disney Studios. They now spend their time between Los Angeles and Western Massachusetts and

---

[2]H. C. Box, *Set Lighting Technician's Handbook: Film Lighting Equipment, Practice, and Electrical Distribution*, 4th ed., Elsevier (Focal Press), New York, 2010.

FIGURE 10.1
Henry Pelham and Eliza Jane.

have two children, Henry Pelham and Eliza Jane (Figure 10.1). Henry plays the piano and at age nine is writing his own music. He enjoys humor and has been drawing and writing his own cartoons since he was four. When the *New Yorker* arrives in the mail, he dives into the cartoons, trying to understand them. Every week he enters the *New Yorker* caption contest (well actually, his mother sends in his entry, due to age restrictions) and is hoping to win.

Eliza, now seven, dotes on the chickens, goats, and other animals they raise on their land. A pig, Stump (he lacks a tail), is the latest to join the menagerie, but he escaped into the woods soon after his arrival. One day Stacey saw him wandering around the yard. She lured him with seed covered in sweetened condensed milk, and soon he was back in captivity. Eliza enjoys music and is learning to play the violin. She also loves to draw, paint, and dance.

Sadly, after many years and two wonderful children, Joan and I parted ways. We have maintained a warm friendship.

CHAPTER ELEVEN

Fisher – Father and Son

I am sometimes asked, "What was Fisher like?" The truth is that while I had a number of contacts with him, I did not know him well. When I married Joan in 1959, Fisher became my father-in-law. We visited him in England, of course, and he came to Madison twice. He died a short time later, in July 1962. When he came to Madison, I recall going with my daughter Helen, then a toddler, to meet him at the airport. He was delighted to see his grandchild, but when he went to take her in his arms, she burst into tears.

Fisher was a scientist versed in many subjects, and as we took our walks in Madison, he impressed me with his deep understanding of the local geology and the flora and fauna. He could be extremely absent minded. While visiting, he lit his pipe while holding a box of matches in his hand, and when the box burst into flames he badly burned his hand.

Someone who *did* know Fisher intimately was his friend, the distinguished geneticist E.B. Ford. After Fisher died, I sent a blank audiotape to Ford asking him to talk about his remembrances. Here is a transcription of part of his response:

> Our [first] meeting, which took place in 1923, was typical of Fisher. Like so many good things in my life, it was due to Julian Huxley . . . . Meeting Fisher somewhere, Huxley mentioned he knew an undergraduate [myself] who had interesting ideas on genetics and evolution. Fisher was a fellow of Caius [at Cambridge]; he was only 33 but was already becoming famous. Other people in his position might possibly have asked briefly about me; a few might even have invited me to go see them. Fisher's reaction was

An Accidental Statistician: The Life and Memories of George E.P. Box, First Edition. George E.P. Box.
© 2013 John Wiley & Sons, Inc. Published 2013 by John Wiley & Sons, Inc.

different. The Fellow of Caius took a train to Oxford to call on the undergraduate!

Characteristically, it did not occur to him to let me know that he was coming, so I was out when he arrived and he settled down in my rooms in College to wait for me. On opening the door of my sitting room on my return, I was surprised to find it full of smoke from pipe tobacco, a thing which disgusts me, and to see a stranger there, a smallish man with red hair, a rather fierce, pointed red beard, and a very white face. ... He took my hand in a firm, bony grip and, bending slightly forward, he gave me a momentary but most searching inspection. Then his face relaxed into a charming smile, the beginning of nearly 40 years of friendship.

...In conversation, I found myself inhibited by some people, and certainly many were inhibited by Fisher, but he and I fit perfectly and, as Roger Knox said, we could tire the sun out with talking. Yet Fisher was not always an easy companion. His vagueness in everyday affairs, and his untidiness, became a legend in its own time, and he could be irritable and inconsiderate. What he wanted took first place and it did not matter too much that other people were inconvenienced. This [was true] on the level of everyday affairs; with things of importance, it was different. He would see his friends through any difficulty or any crisis, sometimes at great trouble to himself.

...He was furious at anything he held to be unjust. I remember his anger when on an important occasion someone mentioned that a candidate for the Royal Society had been the guilty party in a divorce case. Fisher held that nothing whatever but scientific ability and achievement should influence election to the Fellowship. On the occasion I have in mind, the man concerned was certainly not anyone he was supporting but he was not going to have him, as he conceived, unjustly treated.

...Arriving in Cambridge to see him on a summer's day, he said to me, 'You look tired. I shall take you on the river. You shall rest in the canoe while I paddle.' Dear Fisher was half blind, the river crowded, and the canoe unstable. He would charge boat after boat, calling out, 'Look where you're going, Sir!' I cannot think how it was that we were not upset. Since he and I were both in our normal clothes, that would have been most unwelcome. It was a most harrowing experience.

He would take part in anything that was going on. I was present when we all had rounds of pistol shooting after a most excellent picnic lunch party. Fisher with poor sight, his finger on the trigger, waved the gun uncertainly while his friends dived for cover.

...He would leap over intermediate stages in a calculation, leaving his colleagues floundering. I have several times heard a distinguished mathematician say, 'He has evidently solved the problem, correctly, but I don't see how he has done it.' ... He held that mathematics was one way at least of reaching general conclusions but not the only one, and, as he fully recognized, Charles Darwin, whom he so admired, was in mathematics ludicrously incompetent.

Fisher was far ahead of his contemporaries, so far, indeed, that when his epoch-making book, *Statistical Methods for Research Workers*, was published in 1925, it did not receive one favorable review. At the time of his death in 1962, it was in its 14th edition, with reprints, and had been translated into six languages. In 1928 or 1929 he sent a paper for publication to a famous learned society. The assessors turned it down and the officers of the society unwisely left the matter there. It was published elsewhere in 1930 exactly as it stood and it has proved to be one of the fundamental works of evolutionary biology.

... There seems little I can say by way of summary of so great a scientist and so great a friend except perhaps this: he was supremely an individualist, and if ever there was a man whose life was guided wholly by the truth, as he perceived it, it was Sir Ronald Fisher.

Transcription from a tape spoken by E.B. FORD

Fisher had a large family, two boys and seven girls, one of whom died in infancy. The oldest boy, George, joined the air force during WWII and was killed. George was his father's favorite, and Harry, the second son, was a second best. After Fisher died in 1962, Harry lived alone in the family house at Harpenden (Figure 11.1). When I first knew him, I would usually spend a few days there on my visits to England. At that time, Harry and his companions were very keen on rugby football. We would stock up on beer, and until a late hour, various of his friends and their girlfriends would tell stories and sing somewhat improper rugby songs.

Next door lived Mrs. Hester, a mature lady who had been like a second mother to the large Fisher clan. Mrs. Hester was one of those people who could get things done. When she decided that Harpenden needed a swimming pool, for example, she organized the effort to raise money, and thanks to her persuasiveness, the town soon had a very nice facility. The Girl Guides organization (England's equivalent of the Girl Scouts) asked her to be their regional head even though, as she pointed out, she knew nothing whatever about scouting. As always she accomplished the task with spectacular success.

FIGURE 11.1
Harry Fisher.

Children liked Harry immediately. My own children, Helen and Harry, were no different. Later, when Helen was getting married, Harry wrote to her and asked her what she wanted for a wedding present. She said that she would like him to come to her wedding. Harry had never done much traveling, but he flew over the Atlantic first class, and he arrived in the United States in his usual clothes, looking like a tramp. He asked me, "Would these clothes do for the wedding?" I answered with a firm, "No." I took him to an establishment that rented formal wear, where a number of young women were delighted to help him. Harry had a wonderful sense of fun and kept disappearing and reappearing in different guises, to the delight of the young ladies. He was good looking, and in his new attire, you might have mistaken him for an ambassador.

This recalls a story George Barnard told me about Harry's father. Fisher was being honored at a particular ceremony, and George managed to convince him to wear a tuxedo. This Fisher did, but the effect was somewhat spoiled when he appeared in his bedroom slippers.

When Harry died in 2008, his neighbor and friend, Margaret Homewood, wrote about him as follows:

> ... I had known Harry since the 1950s but he became a good friend only about five years ago. I was working in my front garden one day when he came by. He stopped; we chatted about gardening; he offered me courgettes; I offered him a lettuce and we went through to my back garden to cut one ... he was interested in my rough and ready and completely unscientific method of slug control—a border of garlic around the salad bed.
>
> I invited him in to share a pot of tea and we discussed the virtues of the blue Denby jug that held the milk. I said, and it sounds pretentious now as I repeat it, "There is a kind of truth in good design."
>
> And that was it. I'd given the password. My use of the word TRUTH, admitted me to the inner circle.
>
> In the next few hours I discovered that Harry was on a QUEST, a quest for certainty, proof, truth. He would have no truck with metaphysics or mysticism or any religious dogma or doctrine. He could find no basis for belief in God.
>
> He was a student of Science and Mathematics. He looked to Mathematics for truth—that led him to Philosophy—and he found, in his words, "Certainty, both in Philosophy and Mathematics, is currently unavailable."
>
> He found this intolerable and a challenge. Over the last few years he has been hammering out what he called his "proposed System of Deductive Proof in Philosophy." His aim: To put the TRUTH back into Mathematics. To put the TRUTH back into Philosophy," remembering the words of another mathematician: "If the premises are right, so will be the conclusions." He believed his system would challenge all known systems and change the basis of human thought forever.
>
> But now there was a problem, nothing to do with the difficulties of the task.
>
> He was, as I've shown, a man with a purpose, a mission you might say, but that did not include hurting people.
>
> He had no time for bigots and fundamentalists of course, but it was leading respected scholars who were his target: Tarski, Polkinghorne etc.—not your average church-goer, not his family, not his friends, many of whom had firm religious faith.

He was seriously concerned about the impact his findings would have on them.

I used to remind him that there is a long history of religious doubt—even before Richard Dawkins, even before Charles Darwin himself—but still he worried.

He was like no one else—truly eccentric, single-minded—sometimes too direct for comfort, wise and naive at the same time; strangely, unnervingly innocent—childlike.

I was with him—was it really only ten days ago?—when the young consultant registrar, after a long and patient questioning, armed with X-rays, and grey with anxiety, asked Harry's permission to be frank.

"Of course – please."

He was frank—gentle, but totally frank.

There was a pause, then, Harry beamed at the young man, slapped the table twice, saying "Excellent! Excellent!"

. . . I saw it as Harry's recognition of a difficult job well and thoroughly done—step-by-step analysis and then the deduction.

"If the premises are right, so will be the conclusions."

TRUTH, based on evidence.

"I've had a very interesting day," he said, and later, "We've come to the end of the road, haven't we?"

It was, I thought later, another stage of the experiment . . .

This man was now facing mortal illness in the same spirit—a scientist in another phase of the Experiment we call Life—in another area of the great Laboratory—still looking for the TRUTH.

Isn't that a good way to die?

—and a good way to live!

And how glad I am to have known him.

[Citation: With kind permission in a letter to George E.P. Box from Mrs. M.E. Homewood (UK) sometime in 2008.]

CHAPTER TWELVE

Bill Hunter and Some Ideas on Experimental Design

BILL HUNTER arrived in Madison to begin his Ph.D. in the fall of 1960 after working during the summer in Whiting, Indiana. He had enjoyed the job so much that he had written asking whether it was possible to start classes a week after the semester began. I wrote back telling him that no, this would not be possible, so Bill arrived just before classes started (Figure 12.1).

Bill was on a fast track from the day he came to Madison. He completed a brilliant thesis[1] in 1963. Soon after, we offered him an assistant professorship in the Statistics Department. From the beginning, he was an excellent teacher and did some first-class research. As a result, he became an associate professor in 1966, and a full professor in 1969, meaning that he had gone from graduate student to full professor in eight years.

During the first several years I worked at ICI, I taught statistics two nights a week at Salford Technical College to earn a bit of extra income. The college was about halfway between ICI, at Blackley, and my home in Sale, so I would work a full day at ICI, eat dinner at a greasy spoon restaurant, and go directly to the college to teach the class. The course concerned the design of experiments. I wrote out my notes, dittoed them, and circulated them to the class in advance so that the students could listen, rather than write. These notes also formed a basis for later courses

---

[1]W.G. Hunter, *Generation and Analysis of Data in Non-Linear Situations*, Ph.D. dissertation, University of Wisconsin, Madison, 1963.

An Accidental Statistician: The Life and Memories of George E.P. Box, First Edition. George E.P. Box.
© 2013 John Wiley & Sons, Inc. Published 2013 by John Wiley & Sons, Inc.

FIGURE 12.1
Stu Hunter, me, and Bill Hunter.

I taught, and for my contributions to the second book for ICI, called the *Design and Analysis of Industrial Experiments*, otherwise known as "Big Davies," after its editor, published in 1954. The latter book, and my notes, were very useful when Bill Hunter, Stu Hunter, and I began writing *Statistics for Experimenters* in the early 1960s.

As I have said, in the 1950s, statistics was subsumed by mathematics at most universities and received little respect as a field. The incentive to use it in science and industry came from industry itself, not from the college campus. The intention of the book was to introduce its application to a wider public. Stu and I made a good beginning when I moved to Wisconsin in 1959, and he came the following year. In 1962, however, Stu took a job at Princeton and duties at our respective institutions made collaboration more difficult. Stu continued to make further contributions when he became the department's first statistician in residence.

Bill Hunter had a joint appointment with the departments of Statistics and Engineering, and later his office was on the fifth floor of the School of Engineering; you could tell which room was his from outside of the building because he had pasted the pages of the *New York Times* on the window to keep the sun out. We often used Bill's office while we were working on *Statistics for Experimenters*. One of our bigger challenges was that some of our collaboration occurred during the Watergate hearings in the summer of 1973. The hearings, which kept the nation spellbound, regularly lured us from our work to watch the television in the nearby student union. There we were joined by numerous others who found it difficult to resist the unfolding story and its cast of characters. There were the testimonies of John Dean, H.R. Haldeman, and John Ehrlichman, among others, all presided over by the courtly Senator Sam Ervin, a man who possessed remarkable eyebrows.

We continued our work on the book well after the Watergate hearings were over. Finally, in 1978, we sent it off to Wiley, our publisher. Shortly before that, I wrote to my daughter, Helen, who was a senior at Culver. "The house is (almost) unimaginable because over the last several months it has been littered with Galleys, Page Proofs, Indexes and all the etceteras. BE IT KNOWN that today not only has Box (Hunter)^2 been sent off for the last time but 'Fisher The Life of a Scientist' by Joan Fisher Box has also been sent off for the last time. Next time we see them they will be bound books!"

One of the best descriptions of the long process that yielded our book, and of Bill's role in it, comes from Conrad Fung, who was a master's degree student as the book was in progress:

> I met Bill Hunter in the Fall semester of 1975 when I transferred into his course Statistics 424.... It was 'Statistics for Experimenters' taught from mimeographed notes by Box, Hunter, and Hunter, that students could purchase from Mary Arthur in the Statistics Department office. The book would ultimately be published in 1978; we students who continued in the department over the years had the privilege of seeing the book take final form — including the evolution of all the half-normal plots that were in the manuscript when we took the course, into full normal plots when the book was published.

Bill later told the story that he himself had learned from similar notes two decades before, when he got permission . . . to attend George's graduate seminar on experimental design at Princeton. . . . They were dittoed notes in that case, but had the same + and − signs 'marching down the page,' as Bill described them. He said that little did he know then, as a student, that he was learning from a book of which he was to become one of the authors. And upon publication in 1978, he said, he broke into doggerel:

> The three wrote the book page by page
> So that statistics would become all the rage.
> But when it came back from the binding,
> They made a great finding:
> At birth, it had reached voting age![2]

The book would never have become a book had it not been for Bea Shube, the exceptional editor at Wiley Publishing. As a woman working in the field of scientific publishing, she was a pioneer. She started at Wiley in the early 1940s and worked there until 1988, and during that time, she shepherded many outstanding books into publication. There is no question that her encouragement and wise suggestions made *Statistics for Experimenters* a better book. When Bea retired in 1988, I was fortunate that Wiley had Lisa Van Horn to take her place, and she and I worked together for over 20 years, beginning with the 1997 paper edition of *Evolutionary Operation*. In 2004, she oversaw the second edition of *Statistics for Experimenters* and a book of collected articles, *Improving Almost Anything*, in 2006. Lisa also edited the second editions of *Response Surfaces* in 2008 and of *Statistical Control* in 2009. Her input was unusually perceptive, and it was always a joy to work with her. Steve Quigley, the associate publisher at Wiley, oversaw the publication of *all* of these books, and he and I remain close friends.

When I was still in England, I bought some records at bargain prices from a shop that had been damaged by fire. One of these I particularly liked was a song by Cole Porter called "Experiment," sung, I think, by Gertrude Lawrence. I thought it might be an appropriate anthem for our book. But Bill had never heard of it, and no one else he knew in the United States had ever heard of it either. This was strange because

---

[2]C. Fung, "Some Memories of Bill Hunter," Sep. 2009, retrieved from http://williamhunter.net /email/conrad_fung.cfm.

Cole Porter was, of course, American. We knew that if we wanted to use the song in the book, we would have to get permission from whomever owned the copyright. So when Bill was going to England one summer, he decided to track the song down. After several dead ends, he went to the British Institute of Recorded Sound. He asked an old gentleman at the main desk if he knew anything about the song. The man promptly stood up and sang it. When Bill asked why the song was unknown in the United States, the man explained that it was in a 1933 show, *Nymph Errant*, that had been tried out in London and had flopped. So it had never appeared in the United States. In the show, the song is part of a commencement address. We included only the chorus on the first page of *Statistics for Experimenters* II. Here is the song in its entirety:

> Before you leave these portals to meet less fortunate mortals,
> There's just one final message I would give to you.
> You all have learned reliance on the sacred teachings of science
> So I hope through life you never will decline in spite of philistine defiance
> To do what all good scientists do.
>
> *Experiment.*
> *Make it your motto day and night.*
> *Experiment and it will lead you to the light.*
> *The apple on the top of the tree is never too high to achieve,*
> *So take an example from Eve, experiment.*
> *Be curious, though interfering friends may frown,*
> *Get furious at each attempt to hold you down.*
> *If this advice you only employ, the future can offer you infinite joy*
> *And merriment.*
> *Experiment and you'll see.*

[Citation: Cole Porter, "Experiment," from the London stage musical, "Nymph Errant," 1933.]

Statistics for Experimenters is now in its second edition and has sold **over 163,000 copies.** I was particularly delighted when it was translated into Spanish by my friends in Barcelona and Madrid.[3] My friend Ernesto

---

[3] Luis Arimani de Pablos, Daniel Peña Sanchez de Rivera, Javier Tort-Martorell Llabres, and Alberto Prat Bartes worked extremely hard on the first edition, and Xavier Tomas Morer and Ernesto Barrios Zamudio did the same for the second.

Barrios was immensely helpful with the technical revision of the second spanish edition. Ernesto had been one of the longer running Ph.D. students in the Statistics Department. He wrote a very good thesis, but when I encouraged him to take the exam so that he could graduate, he always argued that there were parts of his dissertation that he wanted to improve. He did eventually go back to Mexico, in 2005, and is a professor of statistics at the Instituto Tecnológico Autónomo de México.

In the second edition, we added on the inside covers, more than 60 aphorisms, some our own, and some from other authors. Among them were:

- All models are wrong; some models are useful.
- It's better to solve the right problem approximately than the wrong problem exactly. (John Tukey)
- Experiment and you'll see! (Cole Porter)
- Question assumptions!
- One must try by *doing* the thing; for though you think you know, you have no certainty until you try. (Sophocles)
- Designing an experiment is like gambling with the devil: Only a random strategy can defeat all his betting systems. (R.A. Fisher)
- You can see a lot by just looking. (Yogi Berra)
- Common sense is not common.
- When running an experiment the safest assumption is that unless extraordinary precautions are taken, it will be run incorrectly.
- When Murphy speaks, listen.
- Certain words should be used sparingly. These include should, could, would, ought, might, can't, and won't.

In August 1984, I received a note from Emily Peterson, Chancellor Irving Shain's secretary, asking whether I would be able to join the Chancellor and some special visitors from England for lunch two months hence, on Oct. 18 or 19. I checked my schedule and realized that I was to attend a conference in New Mexico during that time so I wrote back sending my regrets. Ms. Peterson replied immediately, explaining that there was now a chance that the British visitors would be coming to Madison on Friday, Oct. 12 *instead*, and would I be available then? I was free that day, so I answered in the affirmative. In September, Ms. Peterson wrote confirming that I was to join the Chancellor and his English guests at noon on Oct. 12 at "L'Etoile," Madison's finest restaurant, which was across the street from the state capitol building.

On the appointed day, I drove from the west side of town to the restaurant on the capitol square, which is in the center of Madison's famed isthmus. The isthmus is a narrow and congested part of town where parking is very scarce. As I approached the area, traffic became backed up and access to some of the streets was blocked. I suddenly remembered that this was the day Walter Mondale and Geraldine Ferraro were coming to Madison for a huge rally in the run-up to November's presidential election. There was absolutely nowhere to park, and by now I was sure I would be late. Finally I found a parking spot blocks away from the restaurant and hastily made my way to L'Etoile.

I arrived breathless at the restaurant and looked around for Chancellor Shain and his guests. They were nowhere in sight. What I saw instead was a large group of my past graduate students, some of whom had come from great distances. It dawned on me that this was a "setup." Conspiring with the chancellor and his secretary, Bill had invited my students to Madison to celebrate my 65th birthday. I was completely stunned and delighted.

UNIVERSITY OF WISCONSIN-MADISON

CHANCELLOR
Bascom Hall • 500 Lincoln Drive
Madison, Wisconsin 53706
608-262-9946

September 19, 1984

Professor Box:

Chancellor Shain's visitors from England will be here on Friday, October 12. He has arranged for a luncheon meeting with them, you and others on that day at L'Etoile, 25 North Pinckney Street, at 12:00 Noon. I do hope this date is still available on your calendar. Please let us know at your early convenience so that we can confirm a count for lunch to the restaurant.

Emily Peterson
(2-9947)

xc: Professor Box at his home

[Citation: Emily A. Peterson, letter to author, September 19, 1984.]

What behind-the-scenes chaos arose when I canceled the first meeting with the fabricated Englishmen I do not know, but doubtless there was some. Needless to say, there was much merriment and we had a wonderful reunion. Bovas Abraham had helped with the plot, which included asking various students and colleagues, including a number who could not be there, to write letters recounting memories and offering best wishes. These letters were beautifully bound into a leather volume that was presented to me at the gathering. Twenty-five years letter, they still give me great pleasure. The writers were:

Bovas Abraham	Hannes Ledolter
Sig Andersen	Kevin Little
Dave Bacon	Greta Ljung
Steve Bailey	John MacGregor
Don Behnken	Paul Newbold
Gina Chen	Lars Pallesen
Larry Haugh	Dave Pierce
Bill Hill	Jake Sredni
Bill Hunter	David Steinberg
Stu Hunter	Ruey Tsay
Hiro Kanemasu	John Wetz
	Dean Wichern

Below is part of Bill's letter:

My first day in Madison was memorable. I arrived on a Saturday, to register at the last possible moment. I was working that summer in Whiting, Indiana for John Gorman, and did not want to leave. I was having a lot of fun working there, on such things as nonlinear estimation. You came breezing through the department (which, as you recall, was in the house on Johnson Street) about lunch time and asked if I had any plans for lunch. I said no, and you invited me to join you and Gwilym. I then sat in the back of your VW van as you gave him a tour of Madison, which included the zoo.

You stopped at El Rancho for some things for dinner, and, before I knew it, I was having dinner with all of you. There was champagne, which came out of the newly opened bottle in an overly vigorous way, which got everything, tablecloth and all, wet. Napkins were pushed under the tablecloth, which gave the otherwise formal setting, a somewhat casual and definitely lumpy appearance. As I recall, the champagne was a last-minute idea as an addition to the menu, and the bottle was not sufficiently chilled. In any event, a fine time was had by all, and the evening went on to songs by you and Gwilym. I think you both had guitars, and at one point you were singing statistical songs, impromptu efforts in rhyme, with you and Gwilym taking turns.

It was a magical day. About 2 am I left. As I was walking away in the night I thought to myself, 'What a splendid day. It was wonderful. When I tell anyone about it, they won't believe it. Neither will I. I should have a momento. The champagne bottle is a possibility. That would be nice to have.' I turned around, returned, knocked on the door. You looked more than a little surprised when you answered the door, because being called on at 2 am is quite unusual. I explained that I'd like to keep the champagne bottle, and you said fine. That was the end of my first day in Madison.

The best thing about Madison is the friends that I have, which includes Judy, Jack, and Justin. And you, too, George. I love you, and I wish you a happy 65th!

BILL

Soon after coming to Madison, I started an intermediate-level course—Statistics 424—on experimental design. Later Bill taught the course to hundreds of students. One of his requirements was that every student should produce and analyze a factorial design of their own devising and draw appropriate conclusions. One student baked cakes with different ingredients. Another, who was a pilot, experimented with putting his plane into spins and measuring the factors that enabled him to escape them successfully.[4] Below I discuss two examples.

Statistics is about how to use and generate data to solve scientific problems. To do this, familiarity with science and scientific method is essential. In science and technology, it is frequently necessary to study a number of variables. Let's call the variables you can change "inputs," or "factors," and the variables you measure "outputs," or "responses." It used to be believed that the correct way to study such a system affected by a number of factors was by changing one factor at a time. More than 80 years ago, R.A. Fisher showed that this procedure was extremely wasteful of experimental effort. In fact, you should change a number of

---

[4]W.G. Hunter, "101 Ways to Design an Experiment, or Some Ideas About Teaching Design of Experiments," CQPI Technical Report No. 413, June 1975.

factors simultaneously in patterns called "experimental designs." Even now, however, the one factor at a time method is still taught.

Here is a simple factorial design due to Bill Hunter for an experiment on a polymer solution for use in floor waxes with eight experimental runs to study the effects of three factors. These factors were as follows: 1) the *amount* of monomer; 2) the *type* of chain length regulator; and 3) the *amount* of chain length regulator on three responses: milkiness, viscosity, and yellowness.

A 2^3 Factorial Design with Three Responses: Polymer Solution Example

Factor Levels				−	+
1 *amount* of reactive monomer (%)				10	30
2 *type* of chain length regulator				A	B
3 *amount* of chain length regulator				1	3

Factors (Formulations)	1	2	3	Milky?	Viscous?	Yellow?
1	−	−	−	Yes	Yes	No
2	+	−	−	No	Yes	No
3	−	+	−	Yes	Yes	No
4	+	+	−	No	Yes	Slightly
5	−	−	+	Yes	No	No
6	+	−	+	No	No	No
7	−	+	+	Yes	No	No
8	+	+	+	No	No	Slightly

A designed experiment has the merit that it quite often "analyzes itself." For this experiment, with only eight runs, it is clear that milkiness is affected only by factor 1, viscosity only by factor 3, and slight yellowness by a combination of 1 and 2.*

Bill and I believed that it was important to learn by doing. We wanted the class to *experience* how process improvement could be achieved using statistical design. In many of our demonstrations, we used a paper helicopter because it was easy to make, modify, and test. Our basic helicopter design is shown in the figure. In the figure, the heavy lines show where to make cuts in the paper and the dotted lines show where to make folds. If you release the helicopter, it will rotate and fall slowly to the ground. The problem is to modify the design so that the helicopter will stay in the air for the longest possible time.

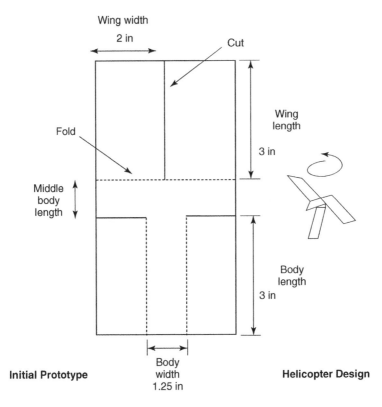

Initial Prototype   **Helicopter Design**

For simplicity we illustrate with eight different helicopter designs arranged in a 2^3 experiment.

Factors			
Wing Length S	Body Length L	Body Width W	Response Time to Fall (in Hundredths of a Second)
−	−	−	236
+	−	−	259
−	+	−	180
+	+	−	246
−	−	+	196
+	−	+	230
−	+	+	168
+	+	+	220

Over these comparatively short flight times, the effects are roughly linear and can be represented by parallel contours plotted on a cube as follows:

The arrow indicates that a helicopter with shorter body (W) width and longer wing (S) length will fly longer, but a change in body length doesn't make much difference. Of course an experiment such as this may be run with more than three factors.

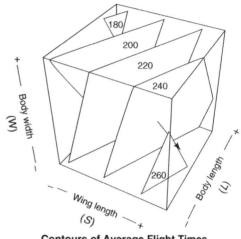

Contours of Average Flight Times

Most of the important ideas in statistics have come about because of scientific necessity, not because of mathematical manipulation.[5] Some examples and the people involved are described below.

A famous example of practice generating theory was Charles Darwin's study of plants and animals on the voyage of the Beagle. Darwin, who was pathetically deficient in mathematics, used *scientific observation* to develop his theory of evolution.

An important factor in evolutionary theory was the *variation* of species. But Francis Galton wondered why this variation did not continually increase. He found that this was because similarities between relatives was only partial, and the partial similarity could be measured by the *correlation coefficient.*

This in turn was taken up with enthusiasm by Karl Pearson, who realized that if we were to find out when items were significantly correlated, it was necessary to discover the *distribution* of the correlation coefficient.

Pearson's methods were clumsy and were developed for large samples. Fisher, however, easily obtained the normal theory distribution and its

---

[5] Parts of this section appeared in the article, "The Importance of Practice in the Development of Statistics," *Technometrics*, Vol. 26, No. 1, Feb. 1984, pp. 1–8.

elaborations using *n-dimensional geometry*. Pearson's methodology also failed to meet the practical needs of W. S. Gosset, when he came to study statistics with Pearson for a year at University College, London, in 1906. Gosset had graduated from Oxford with a degree in chemistry and had gone to work for Guinness's, following the company's policy (begun in 1893!) of recruiting scientists as brewers. He soon found himself faced with analyzing *small* sets of observations coming from the experimental brewery of which he was placed in charge.

Gosset's invention of the *t test* is a milestone in the development of statistics because it showed how account might be taken of the uncertainty in estimated parameters. It thus paved the way for an enormous expansion of the usefulness of statistics, which could now begin to provide answers for agriculture, chemistry, biology, and many other subjects in which small rather than large data samples were the rule.

Fisher, as he always acknowledged, owed a great debt to Gosset, both for providing the initial clue as to how the general problem of small samples might be approached, and for mooting the idea of *statistically designed experiments*.

When Fisher took a job at Rothamsted Agricultural Experimental Station in 1919, he was immediately confronted with a massive set of data on rainfall recorded every day, and of harvested yields every year, for over 60 years. He devised ingenious methods for analyzing these data, but he soon realized that the data that he had, although massive, did not provide much information on the important questions he needed to answer. The outcome was his invention of *experimental design*. Fisher considered the following question: How can experiments be conducted so that they answer the *specific questions* posed by the investigator? One can clearly see his many ideas developing in response to the practical necessities of field experimentation.

Fisher left Rothamsted in 1933 and was succeeded by Yates, who made further important advances. He invented new designs, and showed how to cope when, as sometimes happened, things went wrong and there were missing or *suspect* data.

Later Finney, responding to the frequent practical need to maximize the number of factors studied, introduced *fractional factorial* designs. These designs, together with another broad class developed independently by Plackett and Burman in response to war-time problems, have since proved of great value in industrial experimentation. An isolated example of how such a highly fractionated design could be used for screening out a source of trouble in a spinning machine had been described as early as 1934 by L. H. C. Tippett of the British Cotton Industry Research Association. This arrangement was a 125th *fraction* of a 5^5 design and required only 25 runs!

In another example, Henry Daniels, a statistician at the Wool Industries Research Association from 1935 to 1946, solved the problem of determining how much of the variation in the woolen thread was due to each of a series of processes through which the wool passed. *Variance component* models, which could be used to expose those particular parts of a production process responsible for large variations, had wide application in many other industries.

Later, Henry and I would meet under strange conditions, during the time of the Iron Curtain. We were both members of a small group attending a conference in West Germany who had arranged to travel to the home of Johan Sebastian Bach, at Eisenach in East Germany. At the border, we received a glimpse of what the Iron Curtain involved: an interminable row of concrete dragons' teeth stretching far into the distance, and guards with vicious dogs on leads. The guards took our passports, and they kept them until we crossed back into West Germany. I had not known Henry well before our trip to the border, but I became quite well acquainted with him and his wife during the two hours that we waited to cross into East Germany.

In the development of applied statistics, an important influence was the work of Walter Shewhart on quality control. This work and that on sampling inspection by Harold Dodge heralded more than a half century of statistical innovation, much of it coming from the Bell Telephone Laboratories. This included a rekindling of interest in *data analysis* in a much needed revolution led by John Tukey.

Another innovator guided by practical matters was Frank Wilcoxon, an entomologist turned statistician at the Lederle Labs of the American Cyanamid Company. He said that it was simply the need for quickness that led to his famous Wilcoxon tests, the origins of much subsequent research by mathematical statisticians on nonparametric methods.

An early contribution was by M. S. Bartlett, whose courses I sat in on while I was still at ICI. His work on the theory of *transformation* of data came about because he was concerned with the testing of pesticides and so with data that appeared as frequencies or proportions.

William Beveridge's attempt to analyze time series by fitting sine waves had revealed significant oscillations at strange and inexplicable frequencies. Yule suggested that such series should be represented, not by deterministic functions, but by dynamic systems. Yule's revolutionary idea was the origin of *modern time series models*. Unfortunately, the practical use of these models was for some time hampered by an excessive concern with stationary processes in equilibrium about a fixed mean. Almost all of the series arising in business, economics, and manufacturing do not behave like realizations from a stationary model. Consequently, for lack of anything better, operations research workers led by Holt and Winters devised an non-stationary model. They began in the 1950s to use the *exponentially weighted moving average* of past data and its extensions for forecasting series of this kind. This weighted average was introduced because it seemed sensible for a forecast steadily to discount the past, and it seemed to work reasonably well. However, in 1960, Muth showed that this empirical statistic was an optimal forecast for an important kind of nonstationary model. This model and its generalizations, together with Yule's contributions, later turned out to be extremely valuable for representing many kinds of practically occurring series, including *seasonal series*, and are the basis for so-called ARIMA models.

In further developments, mathematical statisticians had a theory of what they called "most powerful tests," showing that given their assumptions, it was impossible to outperform such a test. In particular, this led to the conclusion that for a binomial testing scheme, you should inspect a fixed number n of items, say 20, drawn at random from the batch, and if the number of duds was greater than some fixed number, say three, you failed the whole batch. Allen Wallis has described the dramatic consequence of a simple query made by a serving officer, "Suppose in such a test it should happen that the *first* three components tested were all duds, why would we need to test the remaining seventeen?" Allen Wallis and Milton Friedman were quick to see the apparent implication that "super-powerful" tests were possible!

At the time Abraham Wald was accepted to be the premier mathematical statistician, but some thought the suggestion that he be invited to work on the problem of a test that was more powerful than a most powerful test was ridiculous. To do better than a most powerful test was impossible! What the mathematicians had failed to see was that the test considered was most powerful only if it was assumed the n was fixed, and what the officer had seen was that n did *not need to be fixed*. This led to the important development of *sequential tests* that could be carried out *graphically*.[6]

A pioneer of graphical techniques of a different kind was Cuthbert Daniel, an industrial consultant who used his wide experience to make many contributions to statistics. An early user of unreplicated and fractionally replicated designs, he was concerned with the practical difficulty of understanding how significant effects could be determined without estimating the size of the experimental error by repetition. In particular he was quick to realize that higher order interactions that were unlikely to occur could be used to estimate experimental error. His introduction of *graphical analysis of factorials* by plotting effects and residuals on probability paper has had major consequences. It has encouraged the development of many other graphical aids, and together with the work of John Tukey, it has contributed to the growing understanding that at the hypothesis *generation* stage of the cycle of discovery, it is the imagination that needs to be stimulated, and that this often best be done by graphical methods.

Obviously one could go on with other examples, but at this point, I should like to draw some interim conclusions.

There are important ingredients leading to statistical advance. They are (1) the presence of an original mind that can perceive and formulate a new problem and move to its solution, and (2) a challenging and active scientific environment for that mind, conducive to discovery.

---

[6]It is heartening that this particular happening even withstood the scientific test of repeatability, for at about the same time and with similar practical inspiration, sequential tests were discovered independently in Great Britain by George Barnard. Nor was this the end of the story. Some years later, Ewan Page, then a student of Frank Anscombe, while considering the problem of finding more efficient quality control charts, was led to the graphical procedure using the sequential idea of plotting the *cumulative sum* of deviations from the target value. The concept was further developed by Barnard who introduced the idea of a V mask to decide when action should be taken. The procedure is similar to a backward-running, two-sided sequential test. Cusum charts have since proved to be of great value in the textile and other industries. In addition, this graphical procedure had proved its worth in the "post mortem" examination of data where it can point to the dates on which the certain critical events may have occurred. This sometimes leads to discovery of the reason for the events.

Gosset at Guinness's; Fisher, Yates, and Finney at Rothamsted; Tippett at the Cotton Research Institute; Youden at the Boyce Thomson Institute (with which organization Wilcoxon and Bliss were also at one time associated); Daniels and Cox at the Wool Industries Research Association; Shewhart, Dodge, Tukey, and Mallows at Bell Labs; Wilcoxon at American Cyanamid; and Cuthbert Daniel in his consulting practice: These are all examples of fortunate conjunctions leading to innovation.

Further examples are Don Rubin's work at the Educational Testing Service, Jerry Friedman's computer intensive methods developed at the Stanford linear accelerator, George Tiao's involvement with environmental problems, Brad Efron's interaction with Stanford Medical School, Gwilym Jenkin's applications of time series analysis in systems applications, and John Nelder's development of statistical computing at Rothamsted.

The message seems clear: A statistician or any scientist who believes himself or herself capable of genuinely original research will find inspiration in a stimulating scientific investigational environment. In all the important scientific developments described earlier, it was the need for new methods in appropriate environments that led to their conception.

As undergraduates, students are encouraged to sit with their mouths open and their teachers pour in "knowledge" for several years. Then those that become graduate students are expected to do something totally different. They have been regularly fed, and now they have to feed themselves and they haven't been taught how to do it. Undergraduate education must provide more opportunities for students to use their creativity—they need help in understanding the art of problem solving. Also, new graduate students tend to start trying to solve a problem in full generality with all the bells and whistles. I've told them, "Don't try to get a general solution all at once. Start out with $n = 1$, and $m = 2$. Once you can *really* understand the problem in its simplest form, then you can begin to generalize." Also, you must try to see the *essence* of the problem. As it says in the New Testament, "Except ye be as little children ye shall not enter the kingdom of heaven."

So I tell my students that it's best if you try to try to think of problems from first principles. It is easy to miss the obvious, and sometimes there is nothing less obvious than the obvious. If you don't approach problems in this way, you may get caught in the tramlines, you think what everyone else has already thought, and you don't arrive at anything new.

Mathematics is primarily concerned with the question: *Given* certain assumptions, is *this* true or isn't it? And in a myriad of other disciplines—physics, chemistry, engineering, and the like—mathematics is an essential tool. But statistics is concerned with finding out things that were *not* in the original model. For example, Albert Einstein noted that many people thought that he developed the idea of relativity from pure theory, but, he said, this was untrue—his theory of relativity was based on observation. Because of the necessity to change the model as one's understanding develops, scientific investigation can never be coherent. One important method that can result in innovation is the interactive process involving induction and deduction. As was said in *Statistics for Experimenters*:

> An initial idea (or model or hypothesis or theory or conjecture) leads by a process of *deduction* to certain necessary consequences that may be compared with data. When consequences and data fail to agree, the discrepancy can lead, by a process called *induction*, to modification of the model. A second cycle in the iteration may thus be initiated. The consequences of the modified model are worked out and again compared with the data (old or newly acquired), which in turn can lead to further modification and gain of knowledge. The data acquiring process may be scientific experimentation, but it could be a walk to the library or a browse in the Internet.

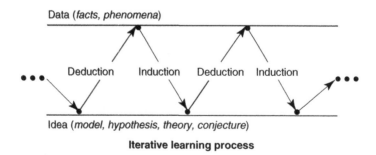

Iterative learning process

The iterative inductive-deductive process, which is geared to the structure of the human brain and has been known since the time of Aristotle, is part of one's everyday experience. For example, a chemical engineer Peter

Minerex parks his car every morning in an allocated parking space. One afternoon after leaving work he is led to follow the following deductive-inductive learning sequence:

Model:	Today is like every other day.
Deduction:	My car will be in its parking place.
Data:	It isn't.
Induction:	Someone must have stolen it.
Model:	My car has been stolen.
Deduction:	My car will not be in the parking lot.
Data:	No. It's over there!
Induction:	Someone took it and brought it back.
Model:	A thief took it and brought it back.
Deduction:	My car will have been broken into.
Data:	It's unharmed and unlocked.
Induction:	Someone who had a key took it.
Model:	My wife used my car.
Deduction:	She probably left a note.
Data:	Yes. Here it is.

Suppose you want to solve a particular problem and initial speculation produces some relevant idea [model, theory]. You will then seek data to further support or refute this theory. This could consist of some of the following: a search of your files and of the Web, a walk to the library, a brainstorming meeting with co-workers and executives, passive observation of a process, or active experimentation. In any case, the facts and data gathered sometimes confirm your conjecture, in which case you have solved your problem. Often, however, it appears that your initial idea is only partly right or perhaps totally wrong. In the latter two cases, the difference between deduction and actuality causes you to keep digging. This can point to a modified or totally different idea and to the reanalysis of your present data or to the generation of new data.

Humans have a two-sided brain specifically designed to carry out such continuing deductive-inductive conversations. While this iterative process can lead to a solution of a problem, you should not expect the nature of the solution, or the route by which it is reached, to be unique.[7]

---

[7]G.E.P. Box, J.S. Hunter, and W.G. Hunter, *Statistics for Experimenters: Design, Innovation and Discovery*, John Wiley and Sons, Hoboken, NJ, 2005.

Your subconscious mind goes on trying to figure things out when you are not aware of it. When something really new occurs to you, it doesn't usually happen when you are sitting down and working at your desk. You might be in the shower or taking a walk, and you suddenly get an idea that solves, or at least helps to solve, the problem. I like working with someone else. The sum of our efforts has always been greater than the parts. Bouncing ideas around with a colleague, discussing, arguing—all catalyze the process of learning and discovery.[8]

Bisgaard[9] defined innovation as the complete process of development and eventual commercialization of new products and services, new methods of production or provision, new methods of transportation or service delivery, new business models, new markets, or new forms of organization. Thus, innovations can occur in marketing, investment, operations, and management techniques as well as in manufacturing and services.

Breakthrough innovation and *incremental innovation* are commonly used terms. Breakthrough innovations are often associated with *new* products or services and incremental innovations with *improvements* in current services or products.

The importance of innovation has sometimes been neglected. Although there are many famous historic examples where innovation was paramount, we pick just one. At the end of World War II, Japanese industry was in ruins. If you visit the Toyota museum in Japan, you will be shown their first car, an exact copy of a Volkswagen. There were many modes of innovation that subsequently brought Toyota cars to world attention. Three of these were (1) previously unknown standards of quality control, (2) new designs developed with the help of thousands of statistically designed experiments, and (3) an attitude toward the workforce based on the idea that they were all one family dedicated to making a good product, with the workers treated fairly by management. Toyota also introduced many other important ideas, such

---

[8]This discussion of innovation is taken in part from the article, "Innovation in Quality Engineering and Statistics," by G.E.P. Box and W. Woodall, *Quality Engineering*, Vol. 21, 2012, pp. 20–29.
[9]S. Bisgaard, "The Future of Quality Technology: From a Manufacturing to a Knowledge Economy and from Defects to Innovations," (2005 Youden Address) *ASQ Statistics Division Newsletter*, Vol. 24, No. 2, 2006, pp. 4–8. Available at http://www.asq.org/statistics/. Reprinted in *Quality Engineering*, Vol. 24, No. 1, 2012, pp. 29–35.

as lean manufacturing. Unfortunately U.S. manufacturers were slow to adopt these concepts. I will go into detail about these facets of Japanese innovation in the next chapter.

The deductive-inductive iteration described earlier is one route to innovation. More generally, some important initiatives that can lead to successful innovation are:

1. Inductive-deductive iteration
2. Lateral thinking
3. Cross-functional discussion
4. Analogy
5. Leadership

In using these ideas, we should not ask which is best but be prepared to employ them all.

With de Bono's lateral thinking, one solves a problem not by working further down the established inductive-deductive path but by finding a new direction.[10] The disadvantage of the inductive-deductive route is that the scheme you arrive at may have already occurred to competing scientists and engineers who have similar education and work with the same set of scientific principles. This is less likely to happen with lateral thinking.

The lateral thinking concept is easier to demonstrate than to define. A simple example of lateral thinking concerns the quandary of a person who has to organize a tennis championship. Supposing that there are 47 contenders, how many matches would be necessary to come up with a winner in a single elimination tournament? The answer could be obtained by enumeration, but it can be reached much more easily by thinking not about the winners but about the losers. There have to be 46 losers, so this is the number of needed contests.

A good statistical example of de Bono's lateral thinking occurred at a Princeton seminar where Merve Muller discussed a way of generating normal deviates by piecewise approximation of the normal curve. This was complicated and messy, and it seemed that there should be a simpler

---

[10]E. de Bono, *Lateral Thinking*, Harper and Row, New York, 1970; and *Lateral Thinking: A Textbook of Creativity*, Viking, New York, 2009.

way. This led me to the following question: "What is it in the normal distribution that is uniformly distributed?" For *two* independent normal deviates, the angle of the radius vector and the log of its length are distributed uniformly and independently. So this provides a way of generating pairs of random numbers. (The less-than-two-page note containing this result[11] has been cited almost 1,400 times on Google Scholar.)

These applications of lateral thinking are not earth shattering, but the idea can be. This was demonstrated, for example, by Charles Darwin. Everyone could see how wonderfully a multitude of living things fitted exactly into our environment, so it seemed obvious that this must be the result of magnificent intelligent design that individually fashioned every living thing and only a super power could accomplish. Darwin, thinking laterally, realized that all that was needed was reproduction and natural selection.[12]

Another example of lateral thinking was R.A. Fisher's use of *n*-dimensional geometry. This led, at once, among other things, to the distribution of the correlation coefficient, to degrees of freedom, orthogonality, the additive property of independent sums of squares, the analysis of variance, the idea of *sufficiency*, the development of regression analysis, and a better understanding of Gauss's method of least squares.

Lateral thinking is counterintuitive and will usually be resisted. It is easy to understand this. We have all been trained to think as if the inductive-deductive mode was the only way to solve problems. Thus, at first Darwin's ideas were contested, as were Fisher's.

The discovery process can also be greatly catalyzed by group discussion, especially if the group contains people from different disciplines. Adair has discussed how teams should be formed and run in order to be most effective.[13] Scholtes et al. described the many aspects that can make this method effective.[14]

---

[11]G.E.P. Box and M.E. Muller, "A Note on the Generation of Random Normal Deviates," *Annals of Mathematical Statistics*, Vol. 29, No. 2, 1958, pp. 610–611.

[12]Alfred Russel Wallace (1823–1913) proposed a theory of evolution based on natural selection independent of Darwin's. Although Darwin has overshadowed Wallace, the two were in regular communication and supported one another in their research.

[13]J. Adair, *Leadership for Innovation: How to Organize Team Creativity and Harvest Ideas*, Kogan Page Limited, London, 1990.

[14]P.R. Scholtes, B.L. Joiner, and B.J. Streibel, *The Team Handbook*, 3rd ed., Oriel Inc., Madison, WI, 2003.

Discussion groups are important not only in themselves but also as a necessary adjunct to the other approaches. Thus, for example, de Bono's "six thinking hats" method can be regarded as either a means to facilitate lateral thinking or as a way to facilitate discussion within groups.[15]

It is important for there to be openness and trust on teams; otherwise, potentially useful ideas may not be suggested. Many point out that with an experienced team, there is little distinction between work and play.

Characteristics of successful teams were evident in the "Monday Night Beer Session" held for years in my home. As noted, students and faculty came from many different departments, and sometimes from industry, to discuss problems and ideas in an atmosphere of open exchange. The sessions were, in the eyes of many who attended, an invaluable learning experience.

Another approach that can prove useful in innovation is the use of analogy. As an example, at ICI one way to improve processes was by running designed experiments, but experiments on the full scale were expensive and disruptive, and small-scale experiments could be misleading. A graphical representation of the imaginary evolution of a species of lobster was used to illustrate to company executives at ICI the idea of *evolutionary operation*. This statistical procedure made it possible to generate information on how to *improve a product during actual manufacture*. Under evolutionary operation, small changes close to normal operating conditions are continually repeated. One is then able to move process factors toward better settings during routine manufacture. This procedure also has the capacity of following moving maxima.

All of these efforts will fail without appropriate leadership. It is true that many people helped Thomas Edison develop the light bulb, many sailors helped Admiral Lord Nelson win the battle of Trafalgar, and no doubt many engineers and scientists helped Steve Jobs develop the iPhone. Nevertheless, these happenings would not have occurred (at least not at that time) without these leaders. One reliable guide to effective leadership is that of Scholtes.[16]

A few years ago, I got a letter from India. The writer said he was a student who very much wanted to study under my guidance. I wrote back

---

[15] E. de Bono, *Six Thinking Hats*, Little Brown and Company, Boston, 1985.
[16] P.R. Scholtes, *The Leader's Handbook: Making Things Happen, Getting Things Done*, McGraw-Hill, New York, 1998.

explaining that I had retired and no longer supervised Ph.D. students. But he responded that that didn't matter as far as he was concerned; he just wanted to be with me. He was given a three-year visa and enrolled in the Industrial Engineering Program at the University. He was quick to learn and very helpful.

Suren had been granted a temporary visa, but I had not realized that he faced a serious problem. He needed to borrow money to pay his fees at the University in dollars, but on an Indian salary, it would take most of his life to pay off. But happily he got a job in the Quality Control Department of the Kohler Company in Wisconsin. The company was very impressed with his work, and they asked the U.S. immigration authorities to provide him with some sort of a document allowing him to stay. Suren went from strength to strength at Kohler, and they used their influence to get him permanent residence, which will allow him to pay off his loan easily.

In 2010, we wrote a paper together introducing a fundamental change in quality control charting.[17] This was published in the journal *Quality Engineering* and received the Brumbaugh Award for the best paper appearing in that year in any of American Society for Quality's journals.

The iteration between practice and theory and the innovation of new ideas is a never-ending process, and sometimes well-established ideas need to be re-thought. An example of this is quality control charts. These were originally developed by Shewhart in the 1930s. The underlying process model had been one where the data was assumed to vary about a fixed mean with deviations that were random. But the fact is that no system behaves in this way. In real processes, the mean and the size and nature of the variation about the mean are not fixed. In our paper, Suren and I pointed out that a more realistic model is provided by the nonstationary integrated moving average (IMA). The reason why this model is of central importance was first explained by John F. Muth in 1960. This leads to an exponentially weighted average quality control chart that can represent reality with much greater closeness.

---

[17] G.E.P. Box, and S. Narasimhan "Rethinking Statistics for Quality Control," *Quality Engineering*, Vol. 22, No. 2, 2010, pp. 60–72.

"The race is over! . . . 'Everybody has won and all must have prizes.'"

CHAPTER THIRTEEN

The Quality Movement

BILL HUNTER and I were very interested in the quality movement from its earliest days, and Bill was closely involved in bringing its techniques to the city of Madison. In 1969, Bill had spent a year in Singapore on a Ford Foundation grant that supplied sophisticated computers and professional expertise to Singapore Polytechnic, where Bill worked with faculty and students. While there, he also worked with another professor to teach an evening seminar on quality techniques for full-time workers in vital positions in Singapore (e.g., those who oversaw the harbor, refuse collection, etc.). He would later teach a similar course in Madison. Bill also traveled to Japan and Taiwan where he visited factories using quality improvement programs.

In the 1970s and 1980s, people in the United States were starting to realize that the Japanese were building cars and other products that were far superior to theirs. This was a spectacular change because before the war, Japanese manufactures had been inferior. Immediately after the Second World War, Japanese industry had been in ruins, and the United States was anxious to help Japan get back on its feet. As part of this effort, two leading experts from the United States, Dr. W. Edwards Deming and Dr. Joseph M. Juran went to Japan to lecture on quality control. Although this subject had largely originated in the West, it had been only sparingly used there. By contrast, in Japan these concepts were taken very seriously and the teachings were spread widely throughout Japanese industry. In

An Accidental Statistician: The Life and Memories of George E.P. Box, First Edition. George E.P. Box.
© 2013 John Wiley & Sons, Inc. Published 2013 by John Wiley & Sons, Inc.

FIGURE 13.1
Bill Hunter and me.

fact, education on the principles of quality control was undertaken as a national project of the highest importance. (Figure 13.1)

But as I have said, there was another factor. In the West, directors and managers seemed to believe that they already knew all there was to know about manufacturing and selling their product. The workers were thought of as disposable underlings who were there to carry out instructions. In particular, this has been the convenient rationale that has justified salaries for upper management that were spectacularly higher than those received by the workers. The Japanese philosophy, on the other hand, was that production was a joint effort in which everyone was personally involved and ideas for improvement were welcomed, rewarded, and celebrated wherever they came from. Their management was paid at a more reasonable level, and the result was that a prodigious number of ideas came from the people who were actually making the product. The methods used for quality improvement were not only those taught by their American mentors, but also those coming from Japanese workers and Japanese experts, such as Professor Kaoru Ishikawa. Because workers developed many new ideas of their own, there was a high level of morale throughout the organization.

One other important concept for quality improvement, which I addressed before, was the use of statistical experimental design conceived in the 1920s by Sir Ronald Fisher for improvements in agriculture. As I mentioned, Fisher showed that it was much better to vary several factors at a time. Fisher's approach was known in Japan as "Taguchi Methods," named for engineering professor Genichi Taguchi. Thousands of designed experiments had been run in Japan to design optimal systems for automobiles.

In 1980, NBC broadcast a special program featuring Dr. Deming that was called *If Japan Can, Why Can't We?* On the show, Deming explained to an American audience that Japan's industrial success in the post-war period relied on statistical methods that would benefit U.S. companies. In Japan, Deming noted, statistical control had led to consistently good quality in a multitude of production processes. Good quality led, in turn, to better control of costs. Moreover, statistical thinking guided *everyone* in the Japanese production process, from line workers to executives.

The effect of these innovations in Japan was dramatic, and they were applied not only to automobiles. Sometime in the middle of the 1980s, I remember seeing a slide that asked, "What do these things have in common?" On the slide were pictures of automobiles, cameras, and every kind of technological gadget. The answer was that for each of these products, the United States had lost 50% of its market to the Japanese in the previous five years. In particular, of course, the U.S. automobile industry had been astonished and embarrassed by the clever designs and narrow tolerances of their Japanese competitors.

Eventually, Deming did get to speak to the top managers of a number of U.S. companies, and a number of senior executives in the Ford Motor Company and other industries visited Japan and discovered that one reason for the superiority of their products was the wide application of experimental design. In particular, they found that Taguchi had used statistically planned experimental arrangements to carry through what he called parameter design. Experimental design had been widely applied in England and in the United States, but only in agriculture. We are indebted to the Japanese for comprehending and demonstrating its enormous value in industrial applications.

In Madison, we started to make careful studies of what was good and not so good about Taguchi's ideas. We examined them in detail and presented a number of research reports. In particular, we found that many of the methods that Taguchi advocated were excellent, but some were not as good as standard procedures that had been developed long ago in the United. Kingdom and the United States. As a result of our studies, we suggested more efficient, and usually simpler, alternatives.

We believed that we needed to inform American industry about these matters and to work on further improvement. To accomplish these goals, in 1985, Bill Hunter and I set up the Center for Quality and Productivity Improvement (CQPI). John Bollinger, Bill Wuerger, and Bob Dye, all from the College of Engineering, worked in various capacities to secure financing and offices for the Center. Bill became the first director, and I was the director of research. In those early days, the Center had an exciting, busy, and even charged atmosphere, and our program assistant, Judy Pagel, was the rudder that kept it on course. We had been given two new assistant professors, Conrad Fung and Søren Bisgaard, who eventually became the second director.

Friendships were strong among those at CQPI, and although we worked hard, there was time for play. The atmosphere was one of a large and happy family. It was about this time that I married Claire Quist. Judy reminded me that often before we were to teach a short course, evey one at CQPI worked very late putting materials together. She reminisced about one occasion when Claire and I brought in ice cream near midnight to cheer everyone up.

Soon after we set up CQPI, Ian Hau came to the department as a graduate student. He was from Hong Kong, where he had been a soccer coach. I heard that he had been using quality techniques to improve a high–school soccer team in the town of Waunakee outside of Madison, and that he had brought the team from near the bottom of its league to near the top. I asked him to give a seminar about this, and his talk was impressive.

I became Ian's thesis supervisor, and after he took his Ph.D., he went to work for a large pharmaceutical company. I expected that he would do standard statistical work, but instead he asked the people at his company what was their biggest problem. They told him that it was the very long time taken to get new drugs approved by the U.S. Food and Drug Administration. After a very careful study of what happened to a specimen drug application, he discovered that much of the delay was due, not to the FDA, but to the company itself. Documents sat on people's desks awaiting attention for long periods of time and similar delays occurred in the necessary testing. Using carefully gathered data, he showed how delays in the system could be dramatically reduced by such simple devices as rerouting and attaching time labels to all documents and procedures. He then went on to solve a number of other strategic problems and in a remarkably short time was promoted to vice president.

Later on my wife Claire and I happened to be in Hong Kong when he was to be married, and I went to the wedding. The Chinese wedding was an interesting experience, but even more was to come when Ian and his wife Grace decided that they wanted also to be married in the United States. Parents play an important part in the Chinese marriage ceremony, and since his own parents were unable to come to the second wedding, Ian asked us to be "deputy parents." It was an elaborate ceremony, and at one point, the bride and groom knelt and presented Claire and I with tea as they would have with their own parents.

When I turned 80 in 1999, I received birthday wishes from Ian in which he reminded me that Claire and I were the only ones who had been to *both* of his weddings. His wife, however, argued that she too had been present on both occasions.

At CQPI we developed many ideas on innovation and quality and we discussed where and when to apply them. We published our work in a new series, the *CQPI Technical Reports*, as well as in the statistical quality literature. We issued the first nine of our reports in February 1986.[1]

---

[1]G.E.P. Box and R.D. Meyer, "Studies in Quality Improvement: Dispersion Effects from Fractional Designs"; G.E.P. Box and R.D. Meyer, "An Analysis for Unreplicated Fractional Factorials"; G.E.P. Box and R.D. Meyer, "Analysis of Unreplicated Factorials Allowing for Possibly Faulty Observations"; W.G. Hunter, "Managing Our Way to Economic Success: Two Untapped Resources"; P.R. Scholtes, "My First Trip to Japan"; B.L. Joiner and P.R. Scholtes, "Total Quality Leadership vs. Management Control"; S. Bisgaard and W.G. Hunter, "Studies

In 1984, Bill Hunter told me that he'd been talking to Joe Sensen-brenner, the mayor of Madison, about how quality ideas could be applied to improve the functioning of the city.[2] The mayor was sympathetic to Bill's ideas and suggested, as a pilot project, to try to improve the running of the city's First Street Garage, which serviced the 900 vehicles belonging to Madison's Motor Equipment Division. He explained in particular that he received numerous complaints from the police about how long it took to get their cars fixed and that there were similar complaints regarding other city vehicles. In addition, worker morale was low and there were tensions between the union and management.

Bill worked on this assignment with Joe Turner, a foreman in the garage, and Terry Holmes, the president of union local 236 (Figure 13.2). They decided to keep a careful record for each vehicle sent in to be fixed of what and when things happened. For example, they recorded how long a vehicle stayed in the back lot waiting to get into the garage, how long it took to get spare parts, how long it took to do the repair, and how long after the repair was completed it stayed on the front lot waiting to be picked up. The data showed where improvement was most needed—by far the longest delay was in waiting for the repaired car to be picked up! Corrective action based on the analysis of data greatly improved the operation of the garage and, as Bill said, gave "employees tools for working smarter, not harder."[3] Once workers learned how to collect and analyze data, there was, in the words of Joe Turner, "more discussing and less cussing."

Another initiative concerned the tree leaf pickup. If you come to Madison, you will be impressed by the number of beautiful trees in the city. Each autumn the collection and disposal of millions of leaves that the trees produce requires a major effort. To get the job done, the city had been divided up into a number of approximately equal areas with a team allotted for leaf collection in each specific area. But by making a

in Quality Improvement: Designing Environmental Regulations"; G.E.P. Box and C.A. Fung, "Studies in Quality Improvement: Minimizing Transmitted Variation by Parameter Design"; W.G. Hunter and A.P. Jaworski, "A Useful Method for Model-Building II: Synthesizing Response Functions from Individual Components."

[2]See also G.E.P. Box, L.W. Joiner, S. Rohan, and F.J. Sensenbrenner, "Quality in the Community: One City's Experience," Center for Quality and Productivity Improvement Technical Report No. 36, June 1989 (originally presented at the 1989 Annual Quality Congress in Toronto).

[3]S. Reynard, "The Deming Way: Management Technique Saves Money in Madison," *The Milwaukee Journal*, March 1, 1985, p. 6.

FIGURE 13.2
Bill Hunter (center) with Joe Turner (right) and Terry Holmes (left) of the First Street Garage in Madison.

survey, Bill found that some of these areas had many trees and others had hardly any, so that some of the teams had almost nothing to do and others were rushed off their feet. Redistributing the areas so that the available people were approximately proportional to the expected amount of leaves produced a much more efficient system. You may say of these problems that the solutions were obvious. That is true, but frequently nothing is less obvious than the obvious. If this were not true, my reputation as a valued consultant would greatly suffer.

From 1972 to 1993, Madison was fortunate to have a remarkable police chief, David Couper. At the time of the anti-Vietnam war demonstrations, his approach was nonconfrontational. When hundreds of students marched down State Street, he walked with them (in ordinary dress, not riot gear). On another occasion, when he was addressing a meeting, a streaker ran across the stage. David merely shook his hand as he ran by.

Chief Couper was very interested in how quality improvement could be applied to the police department. One of the police chief's quality initiatives was directed at slowing down traffic on streets where children were likely to be, and where motorists often drove too fast. He gathered together some parents of young children, and when a speeder was stopped, a parent told the driver about his or her children, and explained how devastating it would be for the parent and for the driver if there were to be an accident.

At the time, it employed Deming's methods in municipal functions, Madison was unique in the United States, but elsewhere in the country there were factories that were beginning to use quality techniques. Bill Hunter had heard about the Motorola television factory near Chicago, which had greatly improved quality control after being taken over by the Japanese. He arranged for his class, members of CQPI, and Terry Holmes and Joe Turner to go down on a bus to take a look. The factory was still being run by the same American workers and management as before, and when we visited, there was not a Japanese person in sight. There were, however, many changes in management policy. In particular, the new philosophy was that "we, the operators, not management, are building this product and we want to take pride in the job we are doing." We learned a great deal about the improvements that had increased quality, many of which had been introduced by members of the workforce themselves. For example, anyone could stop the line if they saw something that was awry, and there was a system of colored tabs that greatly simplified the complicated wiring of the televisions.

When we asked the line managers which system they liked the best, they said that the difference was like night and day. With the old system, almost every television had one or more faults that needed to be fixed *off line*. This was very inefficient, and as one manager noted, "We were running around all day putting out fires." Flaws during production were extremely rare in the new system, as was demonstrated by a large graph hanging from the ceiling that showed a steady decline in the number of defects since the changes in management.

Bill Hunter had taught a course in the evening so that people from the local hospitals, banks, as well as industry could attend. Many of those people became leaders, encouraging quality improvements in their various fields. To coordinate these wider efforts, an organization called

FIGURE 13.3
Trip to Japan.

the Madison Area Quality Improvement Network (MAQIN) was formed in 1987.

In June 1986, a group of us spent two weeks in Japan to observe quality techniques first hand (Figure 13.3). In addition to Jeff Wu and me from Wisconsin, there were Raghu Kackar, Vijay Nair, Madhav Phadke, and Anne Shoemaker from Bell Labs.[4] Bill Hunter would have been invaluable company on this trip, but his fight with cancer was taking all of his strength and traveling was out of the question.

The trip was funded by an NSF grant and Bell Labs, and our visit was greatly facilitated by Professor Genichi Taguchi and Kumiko Taguchi, as well as by Shin Taguchi, who acted as interpreter. Once in Japan, we learned about quality improvement processes, as well as about training and education, at seven Japanese manufacturing companies, and in particular at Toyota. We also visited three trade and professional organizations. Our

---

[4]The six of us wrote "Quality Practices in Japan," in *Quality Progress*, March 1988, pp. 37–41.

goal was to discover how statistical techniques benefited the quality and production processes in Japanese industry.

The period we spent in Japan was enlightening from its very beginning. To our surprise, the trains ran exactly on time and our carriage stopped exactly at a prearranged place indicated on our tickets. We worried when our luggage was immediately taken from us at the station, as we were quite sure that we would have trouble finding it again, but when we arrived at our hotel, we found that each piece had been delivered correctly (to our individual rooms).

When we toured the Toyota plant in Japan, we saw the same system that we had seen in the plant that made televisions outside Chicago. A man from the American auto industry, who was also on the tour, said, "There's nothing new here. This is old technology." He was, of course, looking in the wrong place.

Sadly, Bill's condition weakened throughout 1986. Joe Turner and Terry Holmes had become close friends with "Dr. Bill," as they called him. At his bedside shortly before his death, they asked Bill if there was something they could do for him, and he said, "Yes, you can dig my grave." When the time came, Joe and Terry did just that. Bill died on December 29, 1986 at the age of 49.

In October 1987, the 31st Annual Fall Technical Conference, held in Atlantic City, was dedicated to Bill's memory. Bill's wife, Judy, was there, as were his two sons, Jack and Justin. I had the great privilege of talking about my friend at the conference.

In the late 1980s, Søren Bisgaard, Conrad Fung, and I began to offer week-long short courses for industry to show how the ideas of the quality movement could be put into practice. Some of these short courses were taught on the Madison campus, others were on location at specific industries elsewhere in the country, and some took place in Spain, Sweden, Norway, and Finland. Most of our attendees were people working in industry, in particular, engineers and chemical engineers. With the support of Dean John Bollinger, we charged appropriately high fees for the courses and used the money to fund CQPI.

I met Conrad when he began the Master's program in statistics in 1975. He was among the students who took Bill's Statistics 424 class when it used as its "textbook" the mimeographed manuscript for the not-yet-published *Statistics for Experimenters*. He later became an excellent teacher himself, acknowledging that his method was based on Bill's excellent teaching style.

After completing the Master's degree, Conrad was a statistical consultant for quality control initiatives in manufacturing for DuPont for several years. He then returned to Madison in 1984 to work on a Ph.D. He also took over teaching a statistics course for the American Institute of Chemical Engineers that Bill had always taught. When he completed his Ph.D. dissertation in 1986, he became an assistant professor in the Department of Industrial Engineering, where he and Søren taught courses on statistical methods for quality improvement, stressing real-life applications in industrial settings.

Conrad left in 1992 to run his own statistical consulting business. He never completely left the world of teaching, however, for in addition to running his business, he is also an adjunct professor at the University of Wisconsin Master of Engineering in Professional Practice Program, which provides technical as well as managerial courses for engineers.

Søren Bisgaard and I became acquainted when he was a student in the Department of Industrial Engineering (Figure 13.4). He had had an interesting history. A Dane, he was born in Greenland and had completed an apprenticeship as a machinist before commencing his academic training as an engineer. After getting his Ph.D. at Madison in 1985, he became a valued member of the Center for Quality and Productivity. We got to know each other well and taught many short courses together. He quickly became a professor of industrial engineering, and then he directed CQPI from 1994 to 1998. In 1999, he was the principal founder of the European Network for Business and Industrial Statistics (ENBIS), centered in the Netherlands.

Søren was extraordinarily generous. In 1987, I became ill the day before I was to give a three-hour tutorial at a meeting of the Applied Statistics Network, which was being held in Newark, New Jersey. I was in such pain that I couldn't leave my bed. Søren came over at about 9:30 that morning and sat with me while I explained what I wanted to talk about. He quickly understood it, and by 11:30, he was on his way to the airport to make the presentation for me.

I'm sure that one of the reasons Søren liked Madison was that it is situated on three lakes: Mendota, where the university is located, and Monona and Wingra. Søren and his wife, Sue Ellen, had a beautiful

FIGURE 13.4
Søren Bisgaard and me.

apartment overlooking Lake Mendota, and he kept his boat there. He was a natural sailor and handled his boat like an expert, but you never saw him do it: He would be chatting or opening a bottle of beer and never seemed to be concerned at all with the navigation of the vessel. Claire and I had many happy times sailing with him. Because he had a strong feeling against motorized boats, he was extremely reluctant to use his engine even when we were becalmed in the middle of the lake. On such occasions, it took some time to reach the shore.

Once, when we were in Stockholm together, by some happy circumstance, a friend had lent him his new sailboat. It was a gorgeous boat with fine woodwork and all the latest electronic devices. So we took off among a series of beautiful islands and soon found ourselves sailing alongside another enthusiast. Søren at once initiated a race. The boat seemed to fly along, so close-hauled as to be almost parallel to the water, and with that wonderful smile on his face, he won the race as he had won so many others in his life.

In 1988, the American Society for Quality launched a new journal called *Quality Engineering*. Frank Caplan was the editor, and he asked for my help. This I was happy to provide, so for all the early issues I wrote "George's Column." The series was motivated by the idea that the most important factor in solving problems was common sense. Some titles for this column were as follows:

- *Good Quality Costs Less. How Come?*
- *Changing Management Policy to Improve Quality and Productivity*
- *Teaching Engineers Experimental Design with a Paper Helicopter*
- *Comparisons, Absolute Values, and How I Got to Go to the Folies Bergères*

Later, Søren took over the column, writing articles under the heading, "Quality Quandaries." Many of the articles were co-written with colleagues and were later published in a book called, *Improving Almost Anything: Ideas and Essays* by George Box and friends.

In 2008 Søren was diagnosed with mesothelioma, and it was a tremendous loss when, after a year-long fight, he died in December of 2009 at the age of 58 (Figure 13.5). I think the poem "Sea Fever," by John Masefield, which Claire read at his memorial service, could have been written just for him.

I must go down to the seas again, to the lonely sea and sky,
And all I ask is a tall ship and a star to steer her by,
And the wheel's kick and the wind's song and the white sail's shaking,
And a grey mist on the sea's face, and a grey dawn breaking.

I must go down to the seas again, for the call of the running tide
Is a wild call and a clear call that may not be denied;
And all I ask is a windy day with the white clouds flying,
And the flung spray and the blown spume, and the sea-gulls crying.

I must go down to the seas again, to the vagrant gypsy life,
To the gull's way and the whale's way where the wind's like a whetted knife;
And all I ask is a merry yarn from a laughing fellow-rover
And quiet sleep and a sweet dream when the long trick's over.

FIGURE 13.5
Søren and his wife Sue Ellen shortly before his death.

[CITATION: From SALT-WATER POEMS AND BALLADS, by John Masefield, published by the Macmillan Co., NY, © 1913, p. 55; the poem was first published in SALT-WATER BALLADS, © 1902.]

Our first course, which we called "An Explanation of Taguchi's Contributions to Quality Improvement," was offered in the spring of 1987. This was a discussion of the research that we had done on Taguchi methods in the previous five years. To our surprise, over 60 people took the course. In the CQPI office, Judy Pagel, Conrad Fung, and others often worked well into the night, to get ready for the course. Statistics courses had never garnered this kind of attention in the past. Growing interest in quality improvement methods inspired scientists working in industry to learn all they could about the most effective and efficient statistical procedures.

In the fall of 1987, we began offering our second course, "Designing Industrial Experiments: The Engineer's Key to Quality," which became the short course we taught in Madison as well as "on the road."

The activities of 1988 are suggestive of the enthusiasm that the quality movement engendered at the time. In January, we submitted an NSF proposal for a three-year grant on "Experimental Design and Statistical Methods for the Improvement of Quality and Productivity," which we ultimately received. In April, CQPI brought Professor Noriaki Kano, a Japanese expert in quality management and customer satisfaction, to Madison where he also met with members of the city government and with participants in the Madison Area Quality Improvement Network (MAQIN). In the same month, we held the first of two annual quality retreats at my home, an all-day event that brought together many of the same people from city government and MAQIN. In May, the Annual Quality Congress, held in Dallas, was a "sell out." There Conrad, Søren, Mark Finster, and I gave a tutorial titled "Modern Quality and Productivity Improvement: An Overview" that was well received. In the late summer, Søren, Conrad, and I gave our short course to 60 employees from three Hewlett Packard divisions in Sonoma County, California. In September, we repeated the course for members of the Swedish Association of Engineering Industries in Sodertalje, Sweden. From there I flew to England where I gave a talk at ICI on the uses for statistics in quality improvement. On October 12–14, we brought Professor Kano back to Madison to give two seminars. At the end of October, Stu Hunter, Søren, and I gave a three-part tutorial for the society of Manufacturing Engineers in Chicago.

Because our short courses were successful, we decided in 1990 to make a series of videotapes on the same topics. These were shot at my home in Madison by my son, Harry, who was a cameraman in Hollywood. We filmed two tapes a day for three days. These covered Quality and the Art of Discovery, The Iterative Nature of Scientific Investigation, Factorial Designs, Blocking, Simple Plotting Methods to Analyze Results, and a Practical Demonstration of a Product Development Experiment. The latter featured the same optimization strategy we used in our classes at the university, illustrating the effect of the design optimization of a paper helicopter. Søren climbed a ladder and dropped the different paper helicopters while Conrad sat below timing each flight with a stop watch.

The filming process involved a lot of different people coming and going in our home, because in addition to Harry, there were other crew members who did the lighting, adjunct camera work, and other tasks associated with the taping. At one point, Claire and I decided to escape the chaos by going out to dinner. While in the restaurant, however, Claire had an extreme bout of vertigo. She had recently been diagnosed with Meniere's disease, a disorder of the inner ear that affects balance and hearing, but previously she had never had a major attack. On this particular evening, she was so affected by vertigo that she could barely walk. We left the restaurant, and once in the car, she felt even worse. I pointed the car toward home, driving at a slow, consistent speed while avoiding bumps and potholes in the pavement. We had a number of miles to travel. Suddenly I saw red lights flashing in my rear view mirror and realized that a police car had been following us, so I pulled over. "I stopped you because you're traveling *dangerously* below the speed limit," said the officer, as he inspected us for signs of intoxication. When I explained the situation, the officer gave us free passage home, but once we arrived, Claire crawled into the house and went to bed.

Some years later, Claire had another medical emergency when a drug she was taking caused an extremely painful "bleed-out" in her leg. As I drove her to the emergency room, I once again traveled at a crawl. Luck was not on our side; we were soon pulled over by a police officer who was far from understanding. After taking my license and sitting for a time in his squad car, he returned and said, "You did this before! You have a history of traveling below the speed limit." I was reminded of the story of a fellow who had a car accident. In his written explanation, his insurance agent said, "the policy holder admitted it was his own fault, as he said he'd been run over before."

Our videotapes are still current. They have been converted into DVDs and are once more available.

"What else had you to learn?"
"Well, there was Mystery."

CHAPTER FOURTEEN

Adventures with Claire

...I could write a preface on how we met
So the world would never forget
And the simple secret of the plot
Is just to tell them that I love you a lot...

Lorenz Hart

[CITATION: "I Could Write a Book," by Lorenz Hart and Richard Rodgers, from the musical "Pal Joey;" Broadway stage production 1940; film adaptation 1957.]

CLAIRE and I met when she took a course on quality improvement taught for managers working in and near Madison. At that time, she was the Psychiatric Program Director at Madison General Hospital. She had worked her way through school, earning a Master of Science in Nursing, and had previously taught at the University of Wisconsin School of Nursing. I found that whenever she spoke, it was like a ray of sunshine: helpful, sensible, and original.

Claire told me that becoming a nurse wasn't always easy, and once, as a student, she was nearly removed from the program. She had been working with an older woman who was dying and had no family or friends nearby. On Mother's Day, Claire sent her flowers and made the florist promise to keep her name confidential. The unit clerk called the florist, who revealed Claire's identity as the sender, and there followed many lectures about "inappropriate boundaries." She later had a wonderful

An Accidental Statistician: The Life and Memories of George E.P. Box, First Edition. George E.P. Box.
© 2013 John Wiley & Sons, Inc. Published 2013 by John Wiley & Sons, Inc.

instructor who eventually forgave her and even let her stop wearing her nursing cap because it kept falling off.

When Claire and I were getting to know one another, she sometimes worked evening hours in the hospital's oncology unit, and I would sometimes wait in car with a chilled drink or flowers. We married in September of 1985. On our first anniversary, Claire and I celebrated the day by taking a long walk in the countryside outside of Madison. We hadn't got very far when we heard meowing, followed by the appearance of a thin and scruffy kitten. The cat was obviously hungry and in distress, so we gathered it up and went home. She was a beautiful female kitten with long golden hair, most probably from a litter of barn cats. We couldn't very well call her Anniversary, so she became Annabelle. She later suffered from diabetes, and I took charge of her injections for many years. She was the first of a number of beloved cats, the present one being the large and amiable "Bert."

At the time I married Claire, I had a number of Ph.D. students who gave my new wife a good "looking over." We had purchased a modern house with a swimming pool outside of Madison, and the students helped us move in. The house provided an excellent place for parties, and sometimes when we taught short courses in Madison, we bussed the attendees to our the house for a reception. The students helped by preparing food, serving drinks, and cleaning up. Early in our marriage, one of them, Tim Kramer, approached Claire and told her, "We students have talked and we've decided you're okay for George."

Claire and I had married at a time when the camaraderie among the students and staff at the Center for Quality and Productivity was very strong, and many still speak of the family atmosphere at the Center, where Judy Pagel, who ran the office, was like a second mother to the students. Claire fitted right in and rolled up her sleeves whenever help was needed. She was an excellent organizer and put together many memorable gatherings, some of which were a total surprise. Once, on my birthday, I had a vague notion that she was spending a lot of time cooking, and later in the day, José Ramírez, a student from Venezuela, stopped by. José had a small orchestra called "Paraguas" that played Latin American music, and when eventually other members of this band arrived, it finally dawned on me that my wife was throwing me a party (Figure 14.1).

FIGURE 14.1
Paraguas playing at a birthday party for George, José Ramírez on the left.

One weekend, a contingent of students and professors painted the entire exterior of Judy Pagel's house. Their personalities emerged during the event. Conrad did an expert and meticulous job on all the trim and left each day in clothes that were untouched by a single spatter of paint. Claire worked on the siding, and within minutes, she was saturated in paint from head to toe. Søren mounted the roof with a broom in one hand and a can of insect repellent in the other, cleaning out a bees' nest.

Once, Claire and her students at Edgewood College were having a "punk party." She attired herself in the appropriate outlandish garb, putting several bright colors in her hair, which she stood on end, making her face up appropriately, and attaching safety pins for earrings. Later that day we were having a get-together at the house for the Statistics Department faculty. As a joke I asked Claire to remain in her weird garb. Everyone enjoyed it, but when Norman Draper arrived, her appearance evoked no response, so when I finally asked him what he thought of my wife, he commented, "Well, George, I've liked *all* of your wives."

Claire worked with AIDs patients in the early 1990s, when many were dying. She would often be gone at night to be with someone whose family needed to get a good night's sleep. She told me about one family who kept three televisions on all night long, like a comfort blanket. At about 3 a.m., she turned off the television in the small room where she had been sitting with a dying man. He had not spoken all night, but now he said very clearly, "Don't you like TV?" So that television remained on until she left in the morning. She got to know the man's mother and the pastor and his wife when they visited. When she went to the funeral, she experienced gracious hospitality. The minister's wife came to her, took her hand, and led her to the pew where she herself was sitting.

Claire naturally had many strong feelings and challenges as she did this work, and I remember her writing a lot during this time. It was her way of saying goodbye.

His Bathroom

velvet blue &
gold hand towels
Eternity Calvin Klein
almond massage oil
Jacoma of Paris
Aveda hair spray
Clinique for men
Nivea Oil dry skin
antiseptic soap
paper towels
bleach spray bottle
plastic bags
latex gloves
chux & adult diapers
powder groin rashes
and row after row of
safety capped bottles
soon to be discarded

Claire has an unusual ability to solve difficult interpersonal problems. The shabby way that I had treated my first wife Jessie was something that had sat heavily on my conscience for many years. I told Claire how I felt, and she said at once, "Well you could go and see her and tell her that you are sorry." This had never occurred to me even as a possibility. However,

I wrote to Jessie and she invited Claire and me to come and visit. We flew to Scotland and were received most generously by Jessie, her son Simon, and his wife Wendy.

Before Jessie's death, we were able to make several visits to her in Scotland and to develop a warm relationship with her. This was a transformation I would have regarded as impossible, but Claire understood that the only way to try changing the pain I had caused, and the pain I felt, was to go toward it, not to run away from it.

During our time in Aberdeen, we explored the lovely countryside, watched the salmon running, visited friends, and indulged in a bit of good Scottish Whiskey. In a lovely country home, the owner asked Claire (the only American) if she liked scotch, and she replied, "yes". He returned with a large tumbler and, smiling, handed it to her. She was the driver on that particular trip, but she showed no signs of intoxication. Later she told me that she had used the whiskey to water a number of indoor plants.

Claire and I have a loving relationship. Fortunately we share similar political views and Madison offers many opportunities to express these views. For the most part, however, I don't try to do her things and she doesn't try to do mine. At the First Unitarian Society in Madison, over a period of eight years, she helped conceive a program known as "Quest: A Spiritual Journey." This program provides a two-year spiritual course for its participants and is based on an extensive curriculum that Claire and others wrote. For many, this program has been life changing. Like working with patients who have AIDS, this work is pro bono, but Claire has always said that the rewards are great.

Until quite recently, Claire and I went on many exciting trips together. Some places have become special for us. Claire introduced me to the Point Reyes National Seashore in Marin County, California (Figure 14.2). The park is extremely large—over 70,000 acres—and is set on a peninsula of rugged coastline, beautiful beaches, and high grasslands that are usually almost deserted. In the 1960s, it had been on the brink of destruction by housing developers, but by making it a national park, not only were its spectacular natural features preserved, but so were the ranching and oystering that had provided livelihoods for decades.

In the park we hiked for miles on the extensive trail system. We stayed in the quaint town of Inverness, visiting its pubs, food co-op, and once even a dentist, who in record time repaired my upper bridge so that

FIGURE 14.2
Point Reyes National Seashore.

I could give a lecture the next day without a gap-toothed smile. There were cultural events as well, including a performance of *Under Milkwood* in which the actors used the strangest Welsh accents I had ever heard. Sometimes these were mishaps. Claire had a dramatic reaction to poison oak, for example. But we always returned.

On one occasion, Tom and Helen joined us and we hiked the park together. It was full of all sorts of interesting animals. We spotted a large animal some distance away and began a discussion that continued for quite a period of time. The men were sure it was a mountain lion, and the women were equally sure it was a bobcat. If you asked today, I am pretty sure we would still disagree.

A distinctive feature of the park is the lighthouse shown in the photograph. This was a necessity because the peninsula is one of the foggiest places in North America. Built halfway down a cliff in 1870, it is accessed by a descent of 308 steps. During the San Francisco earthquake of 1906, the peninsula, and, of course, the lighthouse, shifted about 20 feet north, but the lighthouse sustained no damage.

The original inhabitants of the area were the Miwok, who were hunters and gatherers. They were there when Sir Francis Drake appeared in 1579, and the encounter was apparently peaceful. Drake careened his ship, the *Golden Hinde*, in what is now called "Drake's Bay." It is a sheltered inlet where he could repair and restock without interference. A year or two previous to our first visit, a commemorative plaque was erected at a ceremony attended by the British Ambassador. Later the Spanish named the region Punto de los Reyes and established a mission to which they attracted many of the Miwok.

In 1990–1991, I was invited to spend a year at the Institute for Advanced Study in the Behavioral Sciences at Stanford, which gave Claire and me many opportunities to explore the northern coast of California. My colleagues for the year were a very varied and interesting bunch of people. Among them were sociologists, psychologists, behavioral scientists, a chemist, and one other statistician. We each had a hut of our own and were left completely alone to do our research. There was also a basic library and a central staff that would help if called upon.

One intriguing event was the weekly seminar. It took place in the evening and began with hors d'oeuvres and the serving of excellent wine. A requirement was that we each present a talk pitched so that it was intelligible to a person with no specific knowledge of the subject. It was in one of these talks that I learned of the unusual behavior of some fireflies. The speaker, an organic chemist named Jerrold Meinwald, explained that the length of the signal flashes emitted by the female firefly was specific to her variety. He also told us about a very unpleasant species in which the female gave a signal that was not her own, and having attracted the associated type of male, she immediately ate him.

Claire and I loved the islands of the Caribbean. In particular we enjoyed St. Lucia and Barbados, but perhaps our favorite was Dominica. One characteristic of the people of the Caribbean is their interest in the game of cricket. International cricket was played by England, Australia, South Africa, New Zealand, and the West Indies. The West Indies team was made up of players from the previously British Caribbean countries, and the local interest in the game was intense. Everywhere you see little boys with a ball, and a piece of wood for a bat, playing cricket.

We will long remember a taxi ride we took from one end of Dominica to the other. The taxi driver and his friend had a radio and were listening to a running commentary on a game between England and the West Indies in which England was being soundly beaten. The joy of the taxi man and his friend as one disaster followed another would be hard to exaggerate.

Whether it was due to the fresh air or the cricket, the Dominicans were a fine–looking group of people, in marked contrast to some of the tourists who came each week in a large luxury liner. We were surprised to see that most of these passengers never left the ship, although some took pictures from the deck. A few did disembark, but only to walk a couple of hundred yards along the sea front. They had no idea what they were missing.

Perhaps it was because Claire had four Swedish grandparents that we also enjoyed traveling in Scandinavia. In 1987, the Second International Tampere Conference in Statistics was held in Finland, where the waters are a lot cooler than those of the Caribbean. Before the conference, we were met by our friends Katerina and Lars-Erik Öller. We had taken a ship from Stockholm to Helsinki, and Lars-Erik accompanied us as we traveled north on the train to Tampere, and took us to their cottage on the Russian border.

Dr. Tarmo Pukkila ran the conference, and he had arranged that on a free afternoon and evening, we were taken to a lake in a wilderness area 25 kilometers from Tampere, where there were two large saunas, one for the men and one for the women. The lake still had some ice along the edges, having only recently thawed. The idea was that you got cooked in the sauna and then ran outside, dived into the frigid lake, and then ran back. I dipped one toe in the water and that was enough, but most of the men did not venture even that far. Suddenly we saw two women running from their sauna toward the lake, Mrs. Pukkila and Claire. To the amazement of the men watching through the window, they both dived in and swam around in the ice-cold water for some time before they ran back to the sauna. We were even more impressed when someone shouted, "Here they come again!" And sure enough they did—three times in all.

One of the most interesting—and hottest—places Claire and I visited was Egypt. In 1991, the International Statistical Institute met in Cairo, and Claire and I, Bovas and Annamma Abraham, Vijay Nair, Ron Sandland, and several others made the trip. We flew first to Israel, to attend the International Symposium in Industrial Statistics, and in Tel Aviv I had a chance to see my former student, David Steinberg, who was teaching statistics at Tel Aviv University.

When it came time to catch our flight from Tel Aviv to Cairo, we arrived at the airport four hours before our scheduled takeoff because David had told us that getting through Israeli security was a long process. The lines at the airport were at a standstill. This was because each person was minutely questioned and examined. I don't recall how we passed the long wait, but Vijay Nair recalls that I told an "insider's joke" when I said, "I hadn't actually seen a stationary process until now." When we finally had our turn at the security gate, the armed official questioned us closely. We explained that we were traveling to Egypt to present papers at the ISI meeting. He insisted that we each give a few minutes of our presentations, so I took from my briefcase some of the many plastic sheets I used on an overhead projector and gave an impromptu talk.

When we arrived in Cairo, a city of close to 15 million people, we were impressed by the utter chaos in the streets, but Vijay, who had grown up in Malaysia, knew how to handle this. He calmly walked straight into the traffic hand held up. Miraculously vehicles parted to let him through, and we followed.

The disfunction of Egypt's government was again in evidence when we went to the National Museum. This contains famous and precious antiquities, but the experience was sad, for these rare objects were not being cared for. For example, I recalled seeing on television the long queues of people in New York and London waiting to see the Tutankhamen exhibit. We found this in a remote room, where the mask was in a most neglected state.

Later, on a trip to the Valley of the Kings at Luxor, we made our way down a tunnel that was just big enough to crawl through. At the bottom, the walls of the burial chamber were decorated with figures in gorgeous colors. This seemed to be a new tomb only recently opened, and again it was worrying that it was not being properly preserved.

On a moonlight cruise up the Nile, we had dinner on the ship during which we were entertained by a belly dancer. She announced that she would teach someone to belly dance. For some reason I've never understood, she picked on me as a pupil. I think I did pretty well. I know everyone with a camera took many pictures of us (Figure 14.3).

Many members of our group on the Egypt trip will recall how Claire's skills as a nurse were invaluable. Shortly before we left Egypt for the U.S., most of us had taken ill from something we had eaten. Annamma

FIGURE 14.3
Learning to belly dance in Egypt.

Abraham, who was from India, insisted that what would cure us was "hot meat"—in other words, fiery curry. It was our last day in Cairo, and although we had upset stomachs, we found an Asian buffet that served the "remedy." Fortunately Claire had come prepared in the event that the curry failed us. She doled out enough cipro for an army, and we made the long journey home in relative comfort.

It was during this trip that I met my friend Alex Kahlil once again. It had been almost 40 years since I had last seen him in Raleigh. He had returned to Egypt, and rather than pursuing statistics, he had become a citrus farmer.

As I have said, I didn't have to know Claire long to realize that she was a formidable human being (Figure 14.4). A good example of this is when she took on the daunting task of designing and building our house. The first home we purchased had been a wonderful place to live, but it was a fair commute to Madison, so we decided to move closer to town. We bought an older home in the village of Shorewood, about two

FIGURE 14.4
Protest in Madison (Claire and me.)

miles from the university. It was a Cape Cod with an expanded two-story addition at the back, and I had an office on the second floor.

We lived in the house for 6 years, but after a time, climbing stairs became difficult for me. We decided to look for a one-story home, but we found nothing that suited us. Fortunately when we bought the Shorewood house, we had also purchased a lot directly behind it. Claire had planted a garden there, as well as many trees. It occurred to her that, given our poor luck in finding the right house, she might build a new one, and build it on our back lot. I am a professor with a narrow set of skills, so I said, "If you will build it, I promise not to interfere."

Claire set about finding a builder, Associated Housewrights, and working with their architect on a design for a house. She was determined to build the house with environmental concerns in mind, and she found a company that had a similar philosophy. She put a lot of thought into the design, planning the house so that the rooms were both beautiful and accessible. I inhabit the first floor with an office that has every thing I need and a lovely view of the garden. During the construction, she visited the carpenters almost every morning, bearing a plate of cream buns, which they greatly appreciated and which gave her the chance to observe the proceedings. The builders were true artisans. Claire was gone at times during the construction phase because her father was extremely ill and needed her care. On some occasions, the builders had to make decisions without her, and never failed her.

Claire made sure that close by large windows that opened onto the trees, two comfortable chairs sat opposite one another from which there is an enticing view of the back woodland garden. Our view is filled with bird feeders and visited by a multitude of squirrels and raccoons. We spend many hours sitting by the window, in conversation, or in contented silence.

"There's nothing like eating hay when you're feeling faint."

CHAPTER FIFTEEN

The Many Sides of Mac

$\mathcal{F}$OR years, a staunch friend of mine has been Mac Berthouex (or more formally, Professor Paul Mac Berthouex) (Figure 15.1). He taught in the Department of Civil and Environmental Engineering at Wisconsin for 28 years, and he is a world expert on wastewater treatment. His knowledge of how to get a supply of drinkable water to almost any place you care to name has made him internationally famous, and he has spent many years working on projects abroad, usually in poor countries.

Clean water is essential to life, but there is only a certain amount of water in the world, and this must be used and reused. Nature has provided a means for this cleansing to be done. It is achieved by aerobic microorganisms that exist in every body of water that is exposed to air. About ten parts per million of oxygen can dissolve in perfectly clean water, and if you check, you'll find that some quantity slightly less than this is to found in streams, rivers, and oceans. Every natural body of water is to some extent slightly polluted. The pollutants provide aerobic organisms with nutrients, which they absorb at the expense of slightly lowering the level of dissolved oxygen. This sets up a tension, and as more oxygen is needed, more is dissolved, so that we have a permanent system for cleaning up the water supply on the planet. The aerobic organisms are quite remarkable; in a matter of a few hours, they can clean up even raw sewage using the activated sludge process[1] employed in almost every town throughout the industrial world.

---

[1] In older installations, "trickling filters" made use of the same microorganisms.

An Accidental Statistician: The Life and Memories of George E.P. Box, First Edition. George E.P. Box.
© 2013 John Wiley & Sons, Inc. Published 2013 by John Wiley & Sons, Inc.

FIGURE 15.1
Mac Berthouex and me.

Over the years, Mac and Bill Hunter worked on many projects together. Their friendship dated to Mac's student days at Wisconsin in the 1960s. Mac then spent time working overseas in Germany, returning in 1971 to assume a professorship in the Department of Engineering. He and his wife, Sue, moved into University Houses, and soon after their arrival, there was a knock on the door: It was a grinning Bill bearing bread, salt, and wine—"bread, that this house may never know hunger; salt, that life may always have flavor; and wine, that joy and prosperity may reign forever."[2]

Mac and Bill also shared the experience of having lived in Nigeria, although at different times. Mac witnessed the post-colonial disarray in Lagos for several months in 1970, ten years after Nigeria gained its

---

[2]Words that some may recognize from the 1946 Frank Capra film, *It's a Wonderful Life*, although the tradition of giving bread, salt, and wine as housewarming gifts is much older than the movie.

independence from Britain. When he arrived in Lagos, 90% of the city's 800,000 inhabitants used shared taps, wells, and polluted streams for drinking water, and the sanitation infrastructure was virtually nonexistent.[3] Mac's study of the problem, which represented the first time he had used multifactorial experiments on a job, doubled the supply of clean drinking water in the city. Part of his work involved a huge settling tank that had been built by the British in 1922. Two crocodiles lived there, and they would sun themselves on the surrounding wall. Later, when Mac taught experimental design to engineering students at Wisconsin, he proposed a factorial experiment in which one variable was whether there were crocs (or no crocs) in a particular settling tank.

I was also fortunate to work with Mac in the late 1980s and 1990s when he had an NSF grant to study, among other things, how to improve control of wastewater treatment plants. We worked on two sets of data and tried to predict the quality of the effluent coming from the plant from a number of input variables measuring the strength of the effluent and a number of process variables.[4] These predictions would then serve as an early warning of process upsets that could then hopefully be prevented.

Mac's deep knowledge of water quality issues has taken him to Indonesia many times. He was a consultant to the government on environmental management techniques, a visiting professor on two occasions, and helped to design a new engineering campus. One of his biggest projects concerned industrial pollution control in Java. Claire and I were fortunate to benefit from our friend's acquaintance with this beautiful place when he and Sue invited us to join them on a memorable trip to Bali.

I had been to Indonesia under very different circumstances in 1963. At that time, the University of Wisconsin was interested in directing some of its outreach programs toward Indonesia, which was experiencing political tumult. In 1963, Dean H. Edwin Young asked me to go there,

---

[3]M. Gandy, "Planning, Anti-planning and the Infrastructure Crisis Facing Metropolitan Lagos," *Urban Studies*, Vol. 43, No. 2, Feb. 2006, p. 378. This excellent piece may also be accessed on the Internet: http://www.emin.geog.ucl.ac.uk/~mgandy/urbanstudies.pdf.

[4]P.M. Berthouex, G.E.P. Box, and J. Darjatmoko, "Discriminant Upset Analysis," *University of Wisconsin Center for Quality and Productivity Improvement Technical Report* No. 30, May 1988. P.M. Berthouex and G.E.P. Box, "Time Series Models for Forecasting Wastewater Treatment Plant Performance," *Water Research*, Vol. 30, No. 8, Aug. 1996, pp. 1865–1875.

under the auspices of the Ford Foundation, to review a proposal to support a large new university near Bandung. Tensions in Indonesia were high: President Sukarno had increased his ties with Communist China and actively fought the British-sponsored creation of the Federation of Malaysia in September 1963. On September 16, shortly before my arrival, mass demonstrations against the formation of Malaysia had resulted in the burning of the British Embassy and a British Major, Roderick Walker, had defied mob violence by parading up and down in front of the burning embassy, playing his bagpipes in full highland dress.

When I arrived in Jakarta, I was accommodated at the only modern hotel, which the Japanese had built as part of their reparations. At first the huge dining room looked deserted, but from my table in the corner, I became aware that members of the British delegation who had lost their embassy were seated in the far corner across the room.

To facilitate travel, the Ford Foundation allotted me a driver, who, it turned out, had two functions: In addition to driving, he was to carry my handwritten messages, for I was told on no account to use the telephone, which was tapped. I needed to make a number of trips from Jakarta to Bandung. The road was not very wide, and I was surprised one day when my driver guided the car into a ditch. This was to allow the passage of a group of vehicles consisting of two armored cars followed by a Rolls Royce and two more military vehicles, all traveling at great speed on the wrong side of the road. By way of explanation, my driver said, as we got back onto the road, "Presiden!" This was the usual way that Sukarno drove from place to place.

Sukarno liked to be accompanied by beautiful women, and wherever he went there was a phalanx of such ladies lined up beside him. On one occasion, Sukarno was on the other side of the street where my colleague from the Ford Foundation and I were about to cross. Sukarno was heavily guarded by soldiers with a mass of weapons, and it was understood that one was not allowed to cross the street. But my friend did just that, beckoning me to follow. Apparently it was understood that Ford Foundation visitors were immune from attack.

We fell in love with Bali. Mac and Sue found a wonderful place to stay, with simple but comfortable cabanas near the sea. Claire and I would arise at around 6 a.m. each morning, make our way to have coffee at a small restaurant overlooking the sea, and then be off for a walk.

The beach, which was beautiful, had its share of young people selling various items to tourists, always with the assurance that things could be had at a "special price!" Nearby were the shops by the sea. These were canvas structures erected and disassembled each day that sold trinkets and articles of clothing. The ladies who owned the shops soon got to know us, and they were shocked when they realized that Claire and I were prepared to pay the asking price. We got to know Annie, in shop number 10, who was a friend of Mac and Sue's. Annie obligingly offered to teach her how to bargain. And so Claire, who had a Master's degree, got a very different kind of education. Insofar as I could understand it, the critical asking price was about twice the expected selling price, but you needed to proceed by a slow sequence of small adjustments to get there.

Sue, being a teacher of young children, often bought trinkets from Annie to take back as gifts to her students. Customs officials in Milwaukee once suspected Mac and Sue of being covert importers when they saw the packing list detailing what Sue had purchased and shipped to the States. The list contained things like "twenty baskets" and "a dozen dolls," and Mac had a hard time convincing the officials that these were inexpensive mementos for six-year-olds.

The Balinese are remarkable artists, and wood carving is one of their specialties. Master craftsmen patiently teach small groups, composed of perhaps four to six students each. They can be seen seated outdoors around the master. They work extremely slowly and carefully, often taking several weeks to complete a piece. Claire and I bought a beautiful and intricately carved statue of a dancing girl that stands close to four feet tall (Figure 15.2). The statue was made of dense ebony wood and was extremely heavy. We wondered how we were to get this delicately carved work of art some 10,000 miles back to the United States without damage. It turned out that there was a shop in Bali that specialized in packing and dispatching delicate objects in such a way that they were not broken, and sure enough, this was true for our beautiful dancer who today stands in our hallway.

The Balinese also specialize in batik painting, which involves the use of wax and dyes, and we are fortunate to have one of these adorning a wall of our home (Figure 15.3).

Mac has many talents, and these include his ability to write wildly funny plays. On social occasions, he often arrived and, without saying a

FIGURE 15.2
Balinese dancing girl.

word, passed out a play that he had written. Each guest was assigned a role, and the play was acted on the spot.

Mac also writes amusing poems in, most famously, *The Madison Monitor*, a spoof on a local newspaper. One edition of the paper carried the headline, "Berthouex Mistaken for Sociologist," which was accompanied by the following story:

> On Saturday night last, Mac was accosted by a statistician who said, "Who are you? You're a sociologist aren't you?"

FIGURE 15.3
Batik.

Mac admitted that he had wanted to be, but he couldn't pass the practical final examination, which was to get through an open door using only a wood splitting maul and a hand grenade. The door was locked, so he went through the wall. He thought the grade of F was unfair, but the cruelest blow came when the Sociology Examining Committee advised him to go into civil engineering.

He said, "Oh, shit."

"That line of engineering suits you even better," they said.

He was considered a promising sociologist as a result of once asserting at a state meeting that, "You shouldn't worry about whether you are rich or poor so long as you have everything you want."

Dean Bollinger was asked how he felt about his engineering professors being taken for sociologists. He said, "I wish I could get those guys to shave and put on a necktie. That's our biggest problem – that, and the Norwegian chemical engineering professors."

And in a later edition of the *Monitor*, an article titled, "It's Not Cricket":

The English are the only country to have elevated a game to the position of a moral principle. You would never hear an American say, "It's not baseball, old chap." Or a Japanese say, "It's not sumo wrestling."

Robin, it is rumored, during dinner one evening, asked an Englishman to explain the game.[5] Being intoxicated by her charms, and therefore not in his right mind, he agreed and embarked on a lengthy and brilliantly lucid explanation.

He was encouraged in his task by Robin who seemed genuinely interested as he explained the mysteries of silly mid-on, fine leg, googly, chinaman, and so on. At the end of a half-an-hour he sat back, exhausted but satisfied that he had done his bit toward Anglo-America relations by unraveling the mysteries of cricket for a colonial. Robin looked at him for a long time, shaking her head in wonderment, and then said, "That really is remarkable. And to think they do all that on horseback."

[CITATION: Excerpts from the *The Madison Monitor*, a parody of a local newspaper written by, and used with permission from, Paul "Mac" Berthouex.]

For a number of years, Mac and I were members of what we called "The Mashed Potato Club." The club meetings had one item on the agenda: to eat real mashed potatoes. These we found at a local tavern, which, typical for Wisconsin, was family run and had a barroom on one side and a simple restaurant on the other. There was a special each week that featured pot roast, mashed potatoes, and gravy for a pittance. When Mac and I were both in town, we rarely missed this event, and sometimes we permitted other colleagues to indulge alongside us. Unfortunately the tavern closed, but two years ago, we began a new club, "Boys Night

---

[5]"Robin" is Robin Chapman, a good friend, poet, and scientist who is a regular attendee at gatherings.

Out," which includes our friends Brian Joiner and Will Zarwell. At 93, I've become a bit less mobile, and so we have yet another permutation, "Boys Night Out In," which involves getting take-out Thai food and eating it at my house.

"What matters is how far we go? There is another shore, you know, upon the other side."

CHAPTER SIXTEEN

Life in England

Back in 1955, when I was contemplating leaving ICI for a job at Princeton, Cuthbert Daniel asked me in a letter whether I was prepared to live in the United States for the rest of my life. I *have* lived here for the rest of my life, but I have also spent considerable time in England. I still had family there—for some years Jack and Joyce, my brother and sister, were still alive—and made regular visits to see them. I also wanted my children to know their relatives and the country. Although it was usually work that took me to England, there was always time for relaxation, and sometimes I even managed to be a tourist in my own country.

In 1963, after I settled in America, my sister Joyce became ill. I wrote to her doctor, who replied saying there was not anything to worry about. But a week later, I received another letter from the same doctor apologizing because he had confused Joyce with another patient, and that, in fact, she was suffering from incurable cancer and could not be expected to live longer than six months (Figure 16.1).

In the 1960s, there was a strange idea in England that a patient must not be told when she was suffering from an incurable disease. I agreed to this and was able to get away to England for a week. I was allowed to visit only in the afternoons. So I told Joyce that the consulting work I had come to do was a mornings only job. I never knew whether she believed any of this. These were, of course, the very saddest visits I made to England. Joyce was only 52 when she died. She was my good buddy, and I've missed her ever since, and often when I look at my wife, Claire, she reminds one of Joyce.

An Accidental Statistician: The Life and Memories of George E.P. Box, First Edition. George E.P. Box.
© 2013 John Wiley & Sons, Inc. Published 2013 by John Wiley & Sons, Inc.

FIGURE 16.1
Joyce.

Much later in the 1980s, when Claire and I were to visit England, we reserved a room at the Compleat Angler Hotel near Marlow well in advance, and when we arrived at Heathrow, tired and jet lagged in the early morning, it was but a short drive from the airport. Located in a bucolic setting on the Thames, the Compleat Angler was a peaceful and restful place in which to begin our visits to England. The room was always ready for us, and the staff made sure we weren't disturbed.

It was a strange coincidence that Mrs. Hester, who lived next to the Fishers in Harpenden, had a son who had studied hotel management and in due time had been put in charge of the Compleat Angler. He told me a story about Henry Kissinger. Henry had once called the hotel asking that they reserve a suite of six rooms for Mrs. Onassis for a week. Our manager friend explained to him that there were no vacancies and that in any case they had no suites of six rooms. But Henry argued and finally said, "But suppose it were the Palace?" which received the answer, "The Palace would give us proper notice, Sir."

Claire and I made this trip often, and we had a regular route. We always visited Gladys, who lived on the Isle of Sheppey with her daughter and son-in-law, Margaret and Kevin Pender, and their son James. When James was quite young, I sang, "I like bananas because they have no bones," and he would giggle and dance around the table.

There was always a warm welcome from Gladys as she said, "Come in for a cuppa!" A tour of their large and lovely garden followed. The Isle of Sheppey has one of the best fish and chips take-outs in England, so we enjoyed real English food at least once. The time spent with Gladys was delightful. When you went to Gravesend and walked along the boardwalk, there was often someone who would greet her with "Hi Glad!" She had run a sweet and tobacco shop, which became an informal therapy practice for many who visited the shop. During those years, she was called upon to advise about all sorts of issues.

In the vicinity was Brightlingsea, where Mary and George Barnard lived. Here in addition to warmth and tea were Mary's homemade wines, most of which was were apple.

Harry Fisher was next on our journey. He was interested in the genetics of dahlias, and also grew vegetables. He could talk to me about his ideas about mathematics all day, and would have, had Claire not been present.

When we could, we visited Meg Jenkins and her father, Bert. These trips were always anticipated with joy. Claire was given another country to call home. She became an expert driver on the left-hand side of the road.

The lab where I had worked during the war was only eight miles from Salisbury. There is a beautiful cathedral that was completed in 1258 with a tower that is 404 feet tall. On Sundays when I had a day off from the lab, I would take the bus to Salisbury and read my book in the quiet close that surrounded the cathedral.

Years later when I was in England, and Claire flew in to Heathrow to join me, I met her and we drove to Salisbury. She was tired, however, so I left her in the hotel sleeping while I went to the cathedral. To my surprise, there was a big bus there labeled "London Symphony Orchestra." Inside the building, Vladimir Ashkenazy was at the piano rehearsing Rachmaninoff's second piano concerto. The orchestra was accompanying him in their shirtsleeves. I went back to the hotel to get Claire, and when we returned, we sat down next to a man who was the piano tuner. Ashkenazy was hard to satisfy and insisted that the orchestra repeatedly rehearse some of the more difficult sections of the concerto. The piano tuner became more and more nervous since he had to tune the piano before the actual concert began that evening.

I asked if we could get tickets to the performance, but they told me they had been sold out two months ago. We lingered for a bit, and providentially, a man whose robes suggested he was some functionary of the cathedral appeared with two tickets and announced that the previous holder had just died. We bought the tickets and enjoyed the performance.

In 1984, the Royal Statistical Society celebrated its 150th anniversary. Because I had been president of the ASA, I was sent over to represent the organization. We had cocktails in a beautiful building on Whitehall designed by Sir Christopher Wren, and it was there that I met Queen Elizabeth, who had come with Prince Phillip to mark the occasion.

The Queen said, "You don't sound like an American to me." I said, "No! I'm one of Your Majesty's loyal subjects." "You must be part of the brain drain." "More drain than brain, in my case, Your Majesty." She said, "Tell me about Madison. I've never been there." I told her about the lakes and the University and the Madison Symphony Orchestra—in short, we had a lovely conversation. Later people said to me, "What were you talking about with the Queen all that time?" I had not realized that our conversation had been particularly lengthy, because she had a wonderful knack of putting you at your ease—you could almost be talking to your mother.

Claire and I are very fond of tea. In England almost everybody drinks it, so on one of our visits to London, I looked in the telephone book, called the number of a tea importer and said that I wanted to buy some tea. A very refined English voice replied, "Yes, how much tea did you want?" I said, "Oh, about ten pounds." He said, "Ah! Now if you had said ten *tons* I could have accommodated you. But I'll tell you what you do. There's a place near Liverpool Street Railway Station that sells small amounts of tea. It's on Worship Street. I'm sure they will help you."

When we got to Liverpool Street Station, it was raining very heavily. We asked various passersby, who all had umbrellas, where Worship Street was. They all seemed to think it was close but gave contradictory instructions. Because we had no umbrellas, we got very wet, but finally we found it. From the outside, it seemed very rundown—like a set from "Oliver." But we climbed the stairs and we found a large room overflowing with boxes of tea. When we had dried off a little, they asked what sort of tea we wanted. Well we knew the answer to that: We wanted Assam tea. But they said, "What *kind* of Assam tea?" This stumped us, and we asked for guidance. They explained that it was largely a matter of leaf size: The bigger the leaf, the more expensive the tea. They had five different grades, so we settled for ten pounds of medium-grade Assam tea.

After a bit, the boss came in. He had a very broad Scot's accent. I asked him if he liked tea. "No, not at all," he replied. "I'd much sooner have a wee dram o' whiskey." "*Irish* whiskey?" I asked.

Americans drink what I consider a weak sort of tea, and some have been impressed with how strong I prefer mine to be. When I retired in 1990, my fellow countryman Norman Draper wrote the following verse:

A charming young fellow named Box
Who once wrote a paper with Cox
Has aged since that time
And retirement's fine.
Yes, he's been through the school of hard knocks.

So George has retired, we're told.
He hopes to 'come in from the cold.'
But his antics so far
Would certainly bar
A conclusion that he's left the fold.

They seek him in Egypt today.
Tomorrow? In 'Old Mandalay?'
If he's not in Israel
You may pick up his trail
At some future time in UK.

A legendary figure is he.
His students are proud as can be.
If you entertain him
Remember his maxim:
'Two teabags for each cup of tea!'

Even Norm was unaware of my darkest secret: I prefer *three* teabags to two!

CHAPTER SEVENTEEN

Journeys to Scandinavia

$\mathcal{M}$Y connection to Norway began when I met Arnljot and Liv Høyland, who spent the 1987–1988 academic year in Madison. Both statisticians, they attended the Monday Night Beer Sessions, and we became friends. The next year they invited Søren, Conrad, and me to give our "Design of Industrial Experiments" short course in Trondheim. There we met another statistician, John Tyssedal, who told me recently that many who attended the course were surprised by our approach, which was well received. John and I became friends and wrote a paper together.[1]

In Trondheim, we enjoyed sitting by the harbor where a shrimp boat had tied up to the pier and was offering fresh shrimp that had just been cooked on board. It was delicious. A less pleasant sight was a huge bunker where the Germans had hidden their U boats under extremely thick layers of virtually unmovable concrete.

In 1995, I returned to Trondheim to give a course on "The Scientific Context of Quality Improvement." By chance we met a man named Jürgen Ahrend, who was staying at the same hotel. He had restored many historical organs in churches and cathedrals all over the world and was completing the restoration of one of the organs in the historically significant Nidaros Cathedral in Trondheim. The organ was built in 1738–1740 by Johann Joachim Wagner, Bach's contemporary, and a leading organ builder in the late baroque era. During the German

[1] G.E.P. Box and J. Tyssedal, "The Sixteen Run Two-Level Orthogonal Arrays," *Biometrika*, Vol. 83, No. 4, 1996, pp. 950–955.

An Accidental Statistician: The Life and Memories of George E.P. Box, First Edition. George E.P. Box.
© 2013 John Wiley & Sons, Inc. Published 2013 by John Wiley & Sons, Inc.

occupation, the people at the cathedral had hidden their organ—then packed away in pieces—under the church floor. The Nazis eventually discovered it and intended to ship it back to Germany. This, however, never occurred, and today the instrument is the only Wagner organ outside of Germany. It is frequently sought out by concert organists for its full sound.[2] We later attended the cathedral's Sunday service when the refurbished organ was put through its paces by an expert. Claire took in the experience with the eyes and heart of a poet:

Restorations
(Nidaros Domkirke, 1995, Trondheim Norway)

the snow thick, falling in clumps this 30 degree April day
we slosh to the cathedral, a week's celebration for the restored organ
taken to the cellar over sixty years ago, now returned to its place
beneath the dazzling rose window - ten thousand pieces of color

this the only cathedral in Norway, the story is during W.W.II the Germans tried
to remove the organ (made in Berlin 1738) but the town's people
unable to thwart the invasion, hid it and it was left— almost forgotten
until two years ago when a German master began its restoration
today is the service to dedicate that organ, I look around a thousand or more
packed in, speaking Norwegian so I enjoy the aloneness, the freedom from words
inspirited by the light from the windows, the rows of strong faces, children
snuggled in laps, the music of the choir and the spoken yet not heard words

---

[2] As one reviewer said of the organ, "In possession of some pugnacious reeds and a thrilling 'pleno,' it's an ideal choice for ... some big-boned praeludia as well as Buxtehude's most extended organ work: the imposing Fantasia on the Te Deum." Sleeve notes from the recording by organist Christopher Herrick, *Buxtehude: The Complete Organ Works*, Volume 2, played on the organ of Nidaros Cathedral, January, 2009. Recording released January, 2010. Captured from http://www.hyperion-records.co.uk/al.asp?al=CDA67809. See also the remarks on the organ's history by Howard Goodall at http://www.howardgoodall.co.uk/presenting/organsnu.htm.

begin to join in the service, guessing at sounds, knowing when to stand
the melody of the Kyrie, my Swedish Lutheran beginnings are here now, these
solemn faces so like my early churches, the pronunciation irrelevant, my voice
taking part - I am comforted, surrounded by the familiar in an unfamiliar place

stand for the Gospel, watch the sun burst through the windows and the still falling
snow, then sing AMEN, not only sing it but believe it, for I have no memory of
feeling safe in the churches of my youth, they were alive with the pain of my family
and their make-believe words, but here, alone in this cathedral, I am at peace

We went to Stockholm on three occasions and stayed in hotels called, respectively, The Admiral Nelson, The Lady Hamilton, and The Victory. The last two names were, of course, those of Nelson's mistress and of his ship at the battle of Trafalgar. Owned by a couple who had a passion for maritime history, all three hotels were full of Nelson memorabilia. These included newspapers published at the time detailing Nelson's relationship with Lady Hamilton. In the manuscript room of the British Museum, I had been most affected by Nelson's last letter to Lady Hamilton, received after his death at Trafalgar. On it you can see the tear-stained words: "Alas, too late."

Sweden's ties to the sea are everywhere evident. One of her less successful experiences concerns the famous warship, the Vasa. In 1628, a multitude of people assembled at Stockholm's harbor to see the maiden voyage of this brand new ship, which was one of the largest naval vessels of the age. It was 207 feet long and 36 feet wide. It displaced 1,210 tons and had a draught of nearly 15 feet. These dimensions might have been technically appropriate, but unfortunately there had been a last-minute change in plans: The Vasa was originally designed to have four decks, one of which was a gun deck, but the King had heard of a ship built in Denmark that had *two* gun decks, so he insisted that another deck be added and outfitted with 64 bronze guns weighing 71 tons. This interference with the original plan proved disastrous, for as the newly launched ship sailed from the shelter of the tall cliffs, she heeled over and sank.

Fortunately the Baltic Sea has a remarkably low level of salt, and salt is a requirement of the termites that consume wooden wrecks. So when in 1956, the sunken ship was located, having spent 300 years under the sea, it was in remarkably good shape. In 1961, with great caution, it was brought to the surface and transported to a special dockyard.

When we first saw this huge ship, it was still being sprayed intensively with polyethylene glycol to preserve it. On a later visit, the spraying had been completed and the restored ship was on display. More than 700 sculptural and ornamental pieces, some of them quite beautiful, had been recovered.

In Sweden so many people spoke very good English that it was hard to detect those for whom it was not their native language. But Claire and I came across a shop assistant who had a very broad and unmistakable Lancashire accent. So I asked her when she had emigrated from England. "No, I'm Swedish," she said, "but my English teacher was from Lancashire."

"I know something interesting is sure to happen."

Chapter Eighteen

A Second Home in Spain

$\mathcal{W}$HEN I first visited Spain in the 1970s, it was to give short courses that had been organized by Daniel Peña and Albert Prat. George Tiao and I taught time series in Madrid, where Daniel lived, and Stu Hunter and I taught experimental design in Barcelona, the home of Albert.

I had never before realized that Catalonia is a state, for many purposes, separate from Spain, with its own language. Albert was proud of his heritage and was a wonderful guide. He was also an expert chef, and he knew a great deal about wine. Whenever we visited, he took us on a tour of all the restaurants, which while the best, were not necessarily the most expensive. (One place he took us to eat delicious seafood was down a narrow alley where there were a few people who had seemingly passed out on the ground, and I almost feared for my life.)

I remember one day, when we had a free afternoon, he asked me what I wanted to do. I happened to remember that Freixenet, a maker of Catalan champagne, was not far away. When we got there, the gate was closed and it was clearly not a day when they allowed visitors, but Albert found the gatekeeper and chatted with him in Catalan. It was like magic: We were soon shown around with almost royal treatment. On another occasion, Albert's talent as an organizer reached its pinnacle when he put on a conference in Barcelona that featured a dinner at one of Franco's former palaces, replete with strolling musicians.

The early days of my visits to Spain coincided with the end of the Franco regime. Most of our Spanish friends from Madrid and Barcelona had been in trouble with this regime as students. One of these was Agustín

An Accidental Statistician: The Life and Memories of George E.P. Box, First Edition. George E.P. Box.
© 2013 John Wiley & Sons, Inc. Published 2013 by John Wiley & Sons, Inc.

Maravall, who took a time series course that I taught two nights a week. He had come from Spain in 1971 to get his Ph.D. in economics. In the 15-minute break, we used to chat at the coffee machine. He told me that he had been arrested during the "troubles" and sent to a "punishment camp" for a year in Spanish Morocco. The general in charge interviewed Agustín and was happy to discover that he had studied sample surveys. The general told him that very little was known about Spanish Morocco, so he gave him a car and a couple of assistants and they spent the year making a careful survey and census.

Although Franco died on November 20, 1975, his failing health had forced him to give up being prime minister in 1973. That the fascist influence was fading became clear when the regime was openly made fun of on the television.

Franco had surprised his allies in Spain by announcing that Prince Juan Carlos should be his successor and serve as king. When Franco died, Juan Carlos not only assumed the throne but also instituted a democratic government, to the displeasure of the fascists. Tensions grew, and in 1981, fascist forces attempted a coup, bringing machine guns and other weapons into the parliament. Albert told me that he was packed and ready to leave the country at a moment's notice. But the King made a televised appeal for support, and the coup came to a quick end.

Daniel and Albert were wonderful hosts, but Spanish and American ideas about time were different. I remember the first time we were there, in the early 1970s, they invited George Tiao and me to go to dinner. We were ready at about 6 p.m., but they came for us between 9 and 10. We should have known that this would happen because we had sent them the time table for the course as it was taught in the United States, starting at 8 a.m. They wrote back saying, "*You* may start at 8 am, but you'll be the only ones there." So we had to start our courses much later, with a long break for lunch, resuming in late afternoon. People went to bed late, and despite, what were for us unusual hours, they were very serious about learning this stuff. Albert, for example, had about 900 engineering students studying the Design of Experiments at his university.

In the spring of 1986, Claire and I had just flown to Madrid when someone told us that Albert Prat and Tina Roig were getting married that day near Barcelona. We flew at once to join the celebrations that had already begun in a charming outdoor hotel in a park by a lake. There were tables arranged in the garden for the many guests, and much champagne was being consumed. In fact we looked with some surprise at the stack of champagne bottles that must have measured four feet by six. None of it was labeled. I asked Albert how he had got it, but characteristically he only touched a finger to the left side of his nose and said, "I know a man."

The origin of the champagne was not the only mystery because, shortly after we arrived, the bride and groom whispered in our ears, "Don't tell anyone, but we aren't actually married." They had both had previous marriages, one of which had occurred in Germany. They had had a hard time dealing with the various bureaucracies, and at the last minute, there had been some technicality that prevented them from getting a marriage license on that particular day. They decided therefore to go ahead with the wedding celebrations and to get married later. But they said that since their parents were there, they were not telling anybody except for a few close friends.

For the rest of the day, there was merriment with feasting and dancing. I remember that the director of the orchestra wore a toupee that fell down when his directing became too vigorous.

Tina had a large flat in Sitges, a seaside resort close to Barcelona, and we often stayed there on weekends. Sitges is famous for its beaches. It became an artsy and countercultural mecca during the Franco regime, which it remains to this day. We were there once during the "Festa Major," which celebrates the town's patron saint, Bartholomew. The annual celebration features a street procession with outrageously costumed participants, some wearing huge paper maché heads that dwarf their bodies. A dragon breathing real burning coals is dragged through the streets, and firecrackers explode everywhere. At one point, sparks from the dragon burned holes in the back of Claire's dress. Finally, at night, there was the finest fireworks display we had ever seen with rockets fired horizontally over the sea.

When I traveled to Spain in the 1970s and 1980s, Albert always welcomed me to Barcelona with his usual good cheer, and Daniel was an unerringly warm host in Madrid. If my experiences in Spain had ended there, I would have been content. In the 1990s, however, Claire and I

FIGURE 18.1
Albert Luceño and me.

were privileged to live in this beautiful country for two extended periods, and Spain became a second home.

In 1991, Alberto Luceño contacted me saying he would like to visit the department at Madison for a year (Figure 18.1). He was a professor of statistics at the University of Cantabria at Santander, on Spain's northern coast, and he was particularly interested in the work that we were doing on process control. Process control meant different things to statisticians and to control engineers. Statistical process control had been initiated in 1924 by Walter Shewhart, a physicist who was working with engineers at the Western Electric Company in Cicero, Illinois, to improve the quality of telephones. Shewhart's control chart had a line showing the process average with parallel ± limit lines. Data from the process were plotted on a chart. A point outside the limit lines indicated a possible "assignable cause"—a deviation too large to be readily explained by random variation. For such deviations, it was deemed worthwhile to look for the cause, and if it was found, to take action to eliminate it. The idea was that over a

FIGURE 18.2
Marian, me, and Claire.

period of time, common malfunctions would be eliminated. On the other hand, engineering process control automatically *adjusted* the process to stay close to target. A combination of the two ideas is necessary to achieve the best control. Ultimately, Alberto and I published a book about this in 1997, and my good friend, Carmen Paniagua-Quiñones, co-authored the second edition, published by Wiley in 2009.[1]

Alberto and his family did indeed spend a year in Madison, and he and I worked very well together. During this time, Claire and I became close to Alberto and his wife, Marian Ros, who was a physician doing research on in vitro development (Figure 18.2). We also became very fond of their eight-year-old son, Mosqui (short for Mosquito, although he is really named Alberto). This was the beginning of a long friendship, with many visits back and forth between Madison and Santander.

---

[1] *Statistical Control: By Monitoring and Feedback Adjustment*, John Wiley & Sons, New York, 1997. We received the Brumbaugh Award for our paper, "Discrete Proportional-Integral Adjustment and Statistical Process Control," *Journal of Quality Technology*, Vol. 29, No. 3.

We were delighted to see each other again in the summer of 1993, when the International Statistical Institute was holding its meetings in Florence. Claire and I were walking across the Piazza della Signoria when Mosqui spotted us and started running, calling out to us in perfect English, "My mother is pregnant!" If this had been a secret, hundreds of people now knew it. During our time in Italy, Mosqui was a policeman, watching his mother carefully to make sure she didn't drink coffee or do anything else that a pregnant woman "should not" do.

In Italy, we once again saw Agustín Maravall, who was professor of economics at the European University Institute in Florence. He and his family lived in an apartment in the beautiful village of Fiesole, which sits in the green hills above Florence. Agustín recommended that we stay nearby, in the Villa San Giralomo, a pension run by the sisters of the Little Company of Mary. Often called the "Blue Nuns" because of the color of their habit, the sisters of the Little Company of Mary are from an Irish order that had run a hospital and a nursing home in Fiesole. The nursing home had been in the villa, which still housed a small, elderly resident population. The views of Florence from the villa and its gardens were beautiful. Less captivating but still entertaining were the shouts of the plumbers who called to each other through the floors when the plumbing failed during our visit.[2]

We flew from Italy to Spain, where we went to Santander for the first time. Marian and Alberto welcomed us into their home and were wonderful guides to this special region of Spain. Santander is in the Spanish state of Cantabria, near the mountains that stretch across the north of the country and reach down to beautiful beaches. The city lies on an expansive bay, and the location has a rich history. In 1589, a year after the Spanish Armada, Queen Elizabeth received intelligence (false as it turned out) that a number of ships from the remainder of the Armada were being repaired at Santander in readiness for a second try at invading England. So she sent Drake off with a good-sized fleet to take a look. Because the bay is so large, a steady wind blows in from the sea. Drake could not see anything that looked like a threat, and his captains told him that once he entered the bay, he would have great difficulty getting out.

---

[2]For more on the pension and on the Blue Nuns, see L. Inturrisi, "A Monastery Stay: Expect the Austere," *The New York Times*, Oct. 1, 1989.

So they stayed out. Both the history and culture of Santander pulled us in, and we knew that we would be back.

I believe that it was on this same trip to Spain that we had a memorable boating adventure with Xavier Tort. He was my colleague in Barcelona who had helped with the Spanish translation of *Statistics for Experimenters*. He had recently purchased a sailboat and offered to take Claire and I out in it. He raised the sails a bit prematurely, and we started to drift perilously close to an escarpment. Xavier was unable to start the engine, and soon the wind was pushing us against this wall, although we were able to push ourselves away from it. Finally, as things were getting worrisome, he got the engine going and the rest of our trip was uneventful. He recently wrote to me that if the boat had crashed, "I would have become a very famous statistician (my only chance to become so), the statistician that killed George Box!!"

We saw Alberto and Marian again when they spent the summer of 1994 in Madison, the first of five summer visits that allowed Alberto and me to work together. Each time Claire and I would find a house or apartment for them to rent, and in 1994, we also found a crib and other items for their baby, "Peque," then just a few months old (Figure 18.3). These visits were filled with happy times, with many shared meals, trips to the zoo, and spontaneous games to amuse the children.

Early in 1995 I was to receive an honorary doctorate from the University of King Carlos III in Madrid. Before the ceremony, Claire and I stole a few days in Santander to visit Alberto and Marian and their children. They greeted us with dinner and champagne and took us to the small village of Silos to visit the monastery, where we heard the monks singing Gregorian chants. We also went to Haro, in La Rioja, where we visited a very old and large winery. It was our good fortune to have as our guide the owner's granddaughter who was also the great, great, great granddaughter of the founder. She was a graduate of the Department of Viticulture and Enology at the University of California, Davis, and she spoke perfect English. We walked through seemingly endless underground tunnels stacked with wine in barrels all the way. Eventually we entered a room where there was wine more than 100 years old. The room was full of cobwebs. The guide explained that this wine was undrinkable but that the spiders were encouraged because they kept down a species of fly that ate the corks.

FIGURE 18.3
Peque, Mosqui, Claire, George, and Alberto in Madison.

Haro is a little village where everyone is involved with making wine. While there we went to a small restaurant for a lunch of tapas. The man behind the bar came over, took our order, and then asked what we wanted to drink. He was clearly surprised and upset when, after some discussion, we said that we didn't want anything to drink. He returned to his place behind the bar, but he was clearly offended, and after looking at us a number of times, he dived under the bar, came up with a bottle of wine, marched over, dumped it on our table, and spoke loudly and indignantly. Our friends translated this as, "Here, drink this. You can have it for nothing!"

After touring the winery, Claire wrote this poem:

The Winery at Rioja
(January 22, 1995)

the winery López de Heredia, windows
connect the office to the cellar, etched in scenes

from the south of Spain – how she would have loved
these windows I think, she and this place the same age

descend the newly swept stairs with other tourists, down
down into the hand carved cave and thousands of miles
away she died and I once drank this wine and still feel the pain -
the death of her sons, I back away from the guide's racing Spanish

resist translations, chiseled caverns alive
need no explanation – feel the walls, smell the moss, dust thick
the oak and wine invade my senses, swirling with smells
dizzy with the years of richness, I mourn the loss at home

the oak barrels few artisans left to make, the good money
not enough to offset the hard work, what will happen when coopers
are gone and the machine-made barrels feel no care or love

freed to feel the decay of this century, I enter the world of bottles
1/2 million Rioja stacked and never touched, do not disturb
their peacefulness, they have been laid to rest and will enjoy
a second life in 3 or 5 or 7 years – quiet, coated with dust, they wait

last the cementerio, the oldest bottles and Rose the mother
at their final resting, the bottles safely stacked in nichos,
glasses are carefully placed on the lid of a huge discarded cask
and at its center the first vine of this winery, dripping with age
and decay, like the bridal banquet in Dicken's Great Expectations,
yet nearby a 1987 Viño Tondonia awaits its beginning

soon it's tenderly uncorked with instructions about tasting a Rioja
some tourists serious, others snicker and ignore the sober ceremony
– in the wine I taste the oak, the grapes, the decades –
death and life fill my mouth

After a few days in Santander, we all drove together to Madrid where I was to receive the honorary doctorate. Daniel Peña, who had organized this occasion, is a remarkable man (Figure 18.4). In the 1980s, he had become concerned that, although there were two universities in Madrid, there was very little available for poor people. Along with some political colleagues, he persuaded the government to build a new university. Founded in 1989, and housed initially in a former army

FIGURE 18.4
Daniel Peña and me in Spain.

barracks, the new university was named for King Carlos III who had been an enlightened 18th-century ruler who had promoted education and the arts.

The new university has now become a great success with three campuses. It has a student body of approximately 20,000, including graduate students. The main site in Getafe, a suburb of Madrid, has an intimate campus composed of modern buildings arranged around quadrangles. The other campuses are located in Colmenarejo, in the mountains north of Madrid, and in Leganés, just northwest of Getafe. The Scientific and Technological Park at Leganés, which is managed by this university, is the largest in Spain and one of the largest in Europe. It is the commitment to technological innovation that has defined Carlos III since its inception.

There is a considerable international presence here, with seven of the graduate programs taught entirely in English. In a short time, the university had developed three faculties (Law and Social Sciences; Humanities, Communication and Documentation; and the Higher Polytechnic School). These offer over 40 undergraduate as well as doctoral programs.

I have received other honorary doctorates but not like this one. Assisted by our wives, Daniel, the Chancellor, and I got dressed up in splendid gowns with extraordinary beaded hats amid much merriment in the Chancellor's office. At the ceremony, we were led in procession by a line of musicians who played as we entered the hall, where the faculty sat solemnly arrayed in the same finery. Daniel presented me to the Chancellor, who bestowed a number of gifts upon me: white gloves to represent the "purity of my research," a necklace, and a ring. I was also given two massive volumes that were copies of the first edition of Cervantes' *Don Quixote*. These symbolized the gift of knowledge.

As part of the ceremony, Daniel presented the "laudatio," which explained why I was being given the honorary degree. In it, he detailed the usual highlights, but I was especially happy when he said, "One of the most rewarding learning experiences in my life was to attend the Beer and Statistics Seminar that George Box runs in the basement of his home in Madison, Wisconsin. ... In the open and exciting discussion that follows the ... talk I felt as [I have nowhere else] that science is a unique adventure that we approach from different corners [while sharing] a common method and a common perspective: the search for the truth and the understanding of the world and of ourselves."

The ceremony in Madrid gave Claire and me the chance to see a number of dear friends. Albert and Tina came from Barcelona, and we spent happy days in Madrid with them and with Daniel and Mely, who were gracious and generous as ever. The six of us visited the majestic Escorial, at the base of Mt. Abantos, in the town of San Lorenzo de El Escorial, about 28 miles northwest of Madrid. Built of the local gray granite, the Escorial is an impressive compound constructed between 1563 and 1584 by King Phillip II of Spain. Phillip was a Catholic and a deeply religious man, and he dedicated the structure to the glory of God. A few years later, he sent the Armada to England to depose the Protestant, and in his view illegitimate, Queen Elizabeth. The formidable complex houses a monastery, a basilica, Phillip's palace, a pantheon of kings and queens, a library, a museum, extensive gardens, and an art gallery featuring paintings by Titian, Tintoretto, El Greco, Velázquez, José de Ribera, and many others.

We were the only visitors that day, and we were hosted by a Dominican monk, Dr. Agustín Alonso, who was the director of the economics school there. He carried a spectacularly large bunch of keys and promised to show us anything we wanted to see. Dressed in the full habit, he showed us many treasures of this remarkable place. This was followed by a spectacular lunch in his rooms. We had a very happy time together, with good wine, and afterward found ourselves in a spontaneous line dance where we rocked to the tune of *In the Mood* (Figure 18.5).

FIGURE 18.5
Albert Prat, me, and Daniel Peña at El Escorial in Spain.

Having resolved to write a book, Alberto and I needed more time to work together than was allowed by summer visits. In 1995-1996, Claire and I spent a year in Santander, the first of our extended stays in Spain. To learn some Spanish in preparation for our first visit, Claire and I took an evening course at our local community college. After a week or two, it became clear that I was hopeless, and I gave it up. Claire, on the other hand, was a very able pupil. Once we were in Spain, she became competent in the language in a very short time. My Ph.D. student, Ernesto Barrios, who was from Mexico, came with us that year and greatly facilitated our lives because he knew the language.

Alberto and Marian found an attractive apartment for us overlooking the sea front. They also found us a remarkable cleaning lady, Conchi. She turned out to have many talents: She was an excellent cook, she could do the shopping, and she generally looked after our interests. When our toilet leaked, the landlady decided not only to replace it, but also to have the whole bathroom redone. The workmen were slow, very noisy, and smoked excessively. Conchi tore into them like a whirlwind, exclaiming that they were "disturbing the Professor!" She insisted that they not smoke, be quiet, and get on with the job. They complied.

One remarkable feature of Santander is its fishing fleet. Every day except Sunday, they go out into the Atlantic Ocean to catch a wide variety of fish that are sold in a large fish market in the city. The kitchen in our apartment was a challenge because we had just two burners and a microwave, and if we used these simultaneously, a fuse blew. There was a particular fishmonger to whom Claire had explained this situation. He was sympathetic, so on each visit, we would join the queue at his counter, and when our turn came, he would consider the problem of our culinary limitations in the light of the varieties of fish that he had at the time. After careful deliberation, he would make a recommendation and explain exactly how we should proceed to cook the chosen fish using our limited resources. This held up the queue, but those waiting mostly seemed entertained.

The sands at Santander are beautiful, and at the west end of the town, they are exceedingly wide. I've seen two side-by-side football games on the sand at the same time. To the east, however, the beach becomes narrower, and when the tide is in, the sea comes right up to the embankment.

One morning quite early, Claire and I walked down to the east end of the beach. The tide was not fully in, but when we arrived, we saw a good-looking car stranded in the sea. The rule was that no cars were allowed anywhere on the beach. Two youths had apparently ignored this and driven the car onto the sand the previous night. They were frantic. The sea had come in behind them and had cut off any retreat. The only way out was up some very steep steps.

We watched a pantomime that went on for about 45 minutes. First the police came and there was much shouting in Spanish, but no progress was made. Then a small crowd of people assembled, very voluably offering advice. Then a man who was presumably the father of one of the boys appeared. He was quite upset to see his car in a process of slow inundation by the sea. Finally the crowd had to make way for the fire brigade. Ropes were produced, and two stalwarts in galoshes attached them to the car. Eventually the car was drawn sedately by the fire engine through the sea and then pulled with great clanging up each of the steep steps. Once on the road, it was lost in the crowd, so whether it could be driven or what happened to the boys, we never discovered.

I had an office at the University of Cantabria, where Alberto taught. My Spanish never improved much, unfortunately, although communicating with professors and students presented no problem because they spoke English. But there was a woman caretaker in the building who spoke absolutely no English. After a few failed attempts at conversation, I discovered that we had both studied French in school, so we had fun speaking simple phrases to each other—sometimes surprising those around us.

A short distance from where we lived there was a convenient general shop that sold things like cigarettes, bus tickets, and candy. We became friendly with the shopkeeper who told us that he planned to take a trip to England with his wife and daughter. Claire organized a small class in English for him and his family. The man clearly wanted to reimburse Claire for her trouble so he gave her cigarettes—single packs to start with and later entire cartons. Claire didn't smoke but was hesitant to tell him, so she gave the cigarettes to Ernesto, who was a smoker. The cigarettes were American Chesterfields, and Ernesto would have nothing to do with them. His girlfriend wouldn't touch them either, so they were passed on again, and again, until they were finally accepted by someone who must have been desperate for nicotine. Two of Claire's pupils, the wife and daughter, were apt students, but like me, the shopkeeper made no progress learning a new language.

Early on in our stay we had purchased a very nice used English car—a Rover—that had belonged to a researcher who was spending a year in the United States. He had bought it from a garage owner he knew. Rover cars had an excellent reputation, and remarkably, the owner of the garage said that after a year, when we returned the car, he would buy it back from us for almost the same amount we paid for it.

On the morning we were leaving the country, I was waiting with the owner of the garage for Claire, who was bringing the car. After a time I could see a car progressing slowly down the long straight road and from it came great puffs of black smoke and a series of explosions. Quite properly, Claire had stopped at a gas station to refuel the car before returning it. Unfortunately, she had accidentally filled it with diesel fuel.

Although our book was published in 1997, Alberto and I shared many interests. So in 1998–1999, we arranged to live in Santander once again. This time we stayed in an apartment in the Hotel Las Brisas, across the street from the bay. As always, we had wonderful times with our friends, sharing meals and conversation, walking the beach and visiting parts of northern Spain. Claire and Marian were like sisters. They used to meet for coffee, as they were both aficionados of the drink. The first time we lived in Santander, Marian would speak in Spanish and Claire would reply in English. This caused a few looks from others, but it worked well for them.

By this time Peque was four, and even though Claire was making great progress with her Spanish, Peque's Spanish was improving at an even faster pace. When Peque heard something that sounded incorrect, she would say, "No, Claire, you don't say it *that* way." She and Claire had many opportunities to practice, as Claire would sometimes pick her up after school when Marian had to work late, and they would make a trip to a local shop for a supply of candy.

In Spain we celebrated holidays with Marian, Alberto, and the children, and one custom for New Year's Eve that I thought was intriguing was the eating of 12 grapes, one with each of the 12 midnight chimes. This brought good luck, but I never managed to eat the grapes in time, so I will never know if it worked.

FIGURE 18.6
Stacey and me.

FIGURE 18.7
Me, Gaudi, and Harry.

While we were in Spain, we also had visitors from home. Harry and his girlfriend, Stacey, came to stay with us and surprised us by announcing their engagement (Figure 18.6). We enjoyed touring the area with them, and they had an opportunity to explore on their own. We went to Comillas one day, a beautiful village not far from Santander. There we visited "El Capricho," a fanciful and fun summer residence built by Gaudí in 1885–1887 and one of his earliest buildings (Figure 18.7).

We have not been able to return to Spain since our last visit, but we are in close contact with our friends, whose warmth and generosity made Spain a special place for us. Today Mosqui, who is now called "Alberto," is 28 and currently works in the film industry in Toronto. Peque, now "Marian," is a promising student in her first year of college in Barcelona. Their parents are well and deservedly proud of their offspring. Daniel is thriving in Madrid, where he is now rector of his university. Sadly, we lost Albert Prat, who died suddenly on January 1, 2006. He was an excellent statistician and industrial consultant, and moreover, he was a fun-loving man whose generosity knew no bounds.

CHAPTER NINETEEN

The Royal Society of London

$\mathcal{I}$ have been fortunate to receive a number of awards, but I was particularly honored to be elected a fellow of the Royal Society of London in 1985. Next to the Nobel Prize, this is the most distinguished scientific honor you can receive in Great Britain. The Royal Society was inaugurated in 1662 by King Charles II, and since then every British monarch has been a patron of the Society since its inception.

The original document I received from the Society contained stern wording:

21 March 1985
Sir,

We have the honor of acquainting you that you were today elected a fellow of the Royal Society in consequence of which the Statute requires your attendance for admission on or before the fourth Meeting from the day of your election or within such further time as shall be granted by the Society or Council, upon cause shown to either of them, otherwise your election will be void.

You will therefore be pleased to attend at half past four of the clock in the afternoon on one of the following days, cviz:-

Thurs. 28 March   Thurs. 18 April
Thurs. 25 April   Thurs. 2 May

We are,
Sir,
Your Obedient Servants

An Accidental Statistician: The Life and Memories of George E.P. Box, First Edition. George E.P. Box. © 2013 John Wiley & Sons, Inc. Published 2013 by John Wiley & Sons, Inc.

[CITATION: Letter dated 21 March 1985 from the Royal Society of London [UK] to George E.P. Box requesting his presence to accept the position of elected fellow.]

Before the ceremony, the secretary of the society showed us the book that contained the signatures of all the previous fellows (about 8,000 in number). It was humbling to see on these pages the signatures of, for example, Isaac Newton, Charles Darwin, Michael Faraday, and Watson and Crick.

At the ceremony, I was provided with a pen with a nib and an inkwell and I was asked to be careful not to blot the book. At its inception, Charles II gave the society a solid gold mace. In the fellowship ceremony, I took the hand of the president over that mace.

Chapter Twenty

Conclusion

Now a careful reading of *Alice in Wonderland* tells me that I should begin at the beginning, go on to the end, and stop. So if you got that far, you'll recall that I'm color blind, lack finger prints, and am 93. And you've probably guessed that I can't type, nor do I use a computer. None of these facts keep me from wondering what comes next. I'm thinking ny next project will be . . .

An Accidental Statistician: The Life and Memories of George E.P. Box, First Edition. George E.P. Box.
© 2013 John Wiley & Sons, Inc. Published 2013 by John Wiley & Sons, Inc.

Chapter Twenty One

Memories

$\mathcal{H}$ERE are a few excerpts from letters given to me on my 65th birthday in 1984:

> I fondly remember, among others, these "labors of love" over the past years: Proofreading BH², helping Ron Snee with your "Practice and Theory" videotapes, and helping Bill Hunter, Bovas Abraham and Kevin Little with putting together this volume of letters. I wish you the best of health and everything else for the next 2^6 years!
>
> STEVE BAILEY

> P.S. Here's one more "Factional fact": How many persons attended the "Wisconsin dinner" at the ASA Meetings in Philadelphia last month? Why, exactly 64 (or 2^6), of course!

> You stimulated us to learn within the classroom, but like all master teachers, you also provoked us to spend much more time outside of lectures probing and arguing with each other and exploring this unfamiliar territory that held such a special fascination for each of us. From you we learned the importance of regarding published material with polite (and sometimes impolite) skepticism and the reality of the continuing iteration between conjecture and evidence. For those of us who would continue to be involved in research, this was an essential lesson. You also served as an incomparable role model for those of us who would become statistical consultants. The "real world" problems you assigned us, your joint discussion with us concerning these problems, and your restrained but incisive criticisms of our proposed solutions have had a lasting influence. Finally, for those of us whose subsequent careers have involved teaching, your expository skills will be recalled most vividly. The careful presentation, the enthusiastic but sensitively paced presentation, the marvelous use of geometric illustration,

An Accidental Statistician: The Life and Memories of George E.P. Box, First Edition. George E.P. Box.
© 2013 John Wiley & Sons, Inc. Published 2013 by John Wiley & Sons, Inc.

and the ever present good humor were combined to create a classroom experience which we can only try to emulate.

DAVID W. BACON

I'm sitting on my boat anchored in Hope Town Harbor in the Bahamas on a beautiful warm sunny day reminiscing about some of the experiences we shared such as:

—seeing you in action for the first time at an Army Experimental Design Conference in Raleigh during the first week of my graduate studies and then meeting for the first time later that night amid a shower of jokes over a "few" drinks with Sig, Stu, Mike, and Maurice at our apartment.

—the absolutely marvelous summers at Old Bald Smith and Gauss Houses at Princeton with STRG (Statistical Research Group). Never was a PhD thesis started with more enjoyment and stimulation. I probably never thanked you for the opportunity provided but now I do!

DON BEHNKEN

I have many memories of a kind man with many other interests. One who likes Shakespeare plays, Bergman movies, Mahler symphonies, as well as those old radio "Goon Shows," and who does excellent imitations of Peter Sellers. I remember a man sitting at a usually cluttered desk in an office, trying to locate a ringing phone hidden in one of his desk drawers. I remember rollicking Christmas parties at your house where students and professors mixed in an informal atmosphere and could poke fun at each other in skits they had written themselves. The highpoint was usually the song sung by you and Norman Draper. I remember a kind man deeply concerned about social issues. I still have one of your bumper stickers that called for Nixon's impeachment. I remember a distinguished academician who in his colorful Tudor academic regalia walked with me through graduation. It even got us a spot on the local evening news.

Thank you for all these memories.

JOHANNES LEDOLTER

I feel <u>very</u> fortunate that I was able to have you as my advisor. My admiration for you has grown more and more as I have come to realize how much you did for me and all your students. We were able to work on some very interesting and important problems and the end results were very good. Your suggestions were sound and sensible, and they always worked. (Surprisingly often anyway.) I remember being skeptical at times and not always willing to take your advice – only to be proven wrong later on. I am now pleased and proud that our joint work was accepted in some very good journals.

Greta Ljung

Finally, I had the good fortune to have you as my thesis advisor. You taught me sensible ways to attack research questions and forced me to strive to understand clearly the problems before me. This past year, when faced with ideas I didn't understand, I've recalled how you would ask me questions to realize where my understanding was shallow. Your own strong intellect and desire to solve important problems pushed me. Thank you for giving me a sound and exciting intellectual apprenticeship.

Kevin Little

To GEORGE BOX, the greatest teacher I have known:

You taught me an entire philosophy of problem solving and along the way I also learned some statistics. For that I will remain forever grateful.

John F. MacGregor

I was always struck by the amount of attention you paid to the presentation for publication of finished research, though on one occasion this gave me some concern. We had shown that a paper, co-authored by Sir Maurice Kendall, analyzing stock market data, was seriously flawed. I remember that our piece commenting on this paper went through several drafts. Your versions were hard-hitting. I, on the other hand, was nervous, since I was returning to England to begin a career, and was reluctant to develop at such an early stage a reputation for bluntness. Therefore, my versions omitted a lot of exclamation marks and attempted to tone down some of the stronger statements. At the end I thought that all was well. However, it turned out that you had a final go at the galley proofs, so that phrases like "insidious capacity to mislead" crept into the final version. By that state I was amused . . .

Paul Newbold

My first summer in Madison I was assigned to be a Research Assistant to you and Dr. Jenkins who was working with you on the Time Series Analysis book. How patiently you explained to me what Time Series Analysis meant and what you wanted me to do (which I only really comprehended several years later). You gave me a "Mechanical" calculator, you or Dr. Jenkins had brought it from England. It had a hand and if I had to multiply by 6.5, I would move it seven times around and then come back half-a-turn. It was one of my happiest experiences. ... not only what I learned but also walking with you and Dr. Jenkins many times to the lake and having lunch together.

JACOBO "JAKE" SREDNI

One of the most valuable lessons I learned from you is the importance to good statistical research of talking to non-statisticians.

If I could transplant any part of the Wisconsin experience, it would undoubtedly be the Monday night beer and statistics symposium. Those Monday night meetings were a marvelous occasion to see statistics in action ...

DAVID STEINBERG

In my attempt to depict the interactive characteristics that make you the unique person that you are, George, I find that John Maynard Keynes has written the perfect description:

'He must reach a high standard in several different directions and must combine talents not often found together ... He must understand symbols and speak in words. He must contemplate the particular in terms of the general, and touch the abstract and concrete in the same flight of thought. He must study the present in the light of the past for the purposes of the future. No part of man's nature ... must be entirely outside his regard.'

Although Keynes was describing the combination of gifts needed by a master economist, he must have been directing his thoughts toward you, George, the master statistician.

JOHN WETZ

The full text of Bill Hunter's letter that appears in abbreviated form earlier in the book on pp. 164 and 165.

George,

In the fall of 1958, I walked into my first lecture in statistics. Jack Whitwell gave us a reference that day to a paper on playing black jack that had appeared in JASA. We all wrote everything down that day. As the semester progressed, we ceased to write everything down. Some of us began to understand what was important and what was not. Others lost interest in the course and started to guage what was the minimum they could do and still pass the course. I liked the course. Among other things, there were fascinating intellectual puzzles. One that never got answered that semester was, What is a degree of freedom? I know I asked Jack that question during one lecture and after at least two other ones. I never did get an answer that satisfied me. Overall, Jack did a good job of teaching me what he knew, in his gentlemanly and dignified way. In another course that started that semester, a course that lasted the entire year, ~~and in which~~ we — a team of three senior chemical engineering students — were supposed to discover good operating conditions for making phosphoric acid by the clinker process. For us, he put down + and — signs in battle array across one sheet of paper. It was, he told us, a half-fraction of a 2^5 with axial points, in three blocks with replicated center points. We were impressed but did not understand what it was all about or how it was going to help complete the project ahead of us. With his guidance, we set off with a mixed set of feelings — confusion, hope, and some optimism. Although there were many things we did not know, we did know something was wrong when, in our first block of ten runs, the highest and lowest values were two replicated center points.

When we had started we thought that a reasonable range for the ~~temperature~~ time was two to five minutes. What we realized was that, unfortunately, the reaction was much faster than we had anticipated, and it had gone to completion before two minutes. So, with Jack's help, we started all over again, this time with a ~~temperature~~ time range of 15 seconds to 60 seconds. Things worked out much better this time. We completed all the blocks, and we analyzed the data. We even did a canonical analysis, which we regarded as fairly mysterious business. Only two canonical variables were needed to explain all the data we had collected that year. We were suitably impressed with the power of statistical methods. The response surface we showed in our final report looked like a car muffler.

When I told Jack I wanted to take another course in statistics in my final semester at Princeton, he suggested that I take a special graduate course you were going to be giving. Your course was going to be given in the Chemical Engineering Department. Being only an undergraduate, I had to go on a scavenger hunt and collect signatures of five deans before being allowed to enroll. I was the only undergraduate in that course. I remember being given damp, limp pages reeking of ditto fluid at the beginning of many lectures. There was great excitement in the air. These notes, we were told, would be turned into a book. Little did I suspect, in that room with the high ceilings, big windows, and creaky wooden floors, in that building with twelve-foot thick stone walls, that I would end up being a co-author of that book.

I thoroughly enjoyed that course. Maybe I shouldn't say "thoroughly" because it was a strange situation to be surrounded by graduate students and visitors from outside the university, from such places as Fort Monmouth, and I didn't know exactly where the course was headed (all other courses I had taken had a textbook that was finished, in hard covers, in the book store), and I didn't know what was expected of me. In the first homework assignment I got mixed up between a paired and an unpaired comparison. I learned a great deal in that course. I was enthralled with the geometrical explanations of things. I finally learned what degrees of freedom were. I saw how statistics could help people learn from data — both in collecting good data in the first place and in analyzing them. I decided I wanted to learn more, and I discovered somehow — I don't remember how — that you were headed to Wisconsin to start a statistics department there, and that it wouldn't accept students until 1960. I made plans to go to the University of Illinois and get a master's degree in chemical engineering. My intent was then to go to Wisconsin. Your lectures did not stop that second semester when everyone else's did. Someone in Nassau Hall had set up an academic calendar, but you were marching to a different drummer. The graduate students weren't going anywhere. The visitors wanted to keep visiting and learning what you had to say. I kept coming to lectures because I was sticking around for graduation ceremonies. On those lovely spring afternoons in 1959, I would be playing frisbee with friends. I'd say I had to leave and go to a lecture. They would say that I couldn't be going to lectures because they were all finished. "Not mine," I would say and go off to learn about fractional factorial designs. I graduated and left, but your lectures continue on into the summer without me. I never have learned when they ended. Before I left I took an exam, an oral exam. I answered all the questions you asked, except the last, which you prefaced with the words, "I don't expect you to be able to answer this one." It had something to do with a 2^{8-4} design. You asked me, at the end of the exam, what I intended to do after I left Princeton, and I told you I wanted to come to Wisconsin in 1960.

I did. And I'm still here. There are times when I realize just how fortunate I am to have ended up in Madison. One of those times was when we were heading back here after our year in Africa and travels associated with that trip. (It was then that I came to three conclusions: The rate of population growth should not continue at its present level; you can get pretty good beer just about anywhere; and Madison was a very good place to live.) Friends have been important to us. I'm happy that I've gotten to know you better. To see how you have enjoyed so many things — big and small, to see the pleasure you have experienced with Helen and Harry, to have been able to share good times and bad times with you — these have all meant very much to me. Another time when I have realized how lucky am I to have ended up in Madison was the other day when I played golf in Spring Green (first time I had played in

fourteen years) on a beautiful fall day. There are
so many things I like about Madison. One is
certainly the physical surroundings, the countryside,
the rivers, the sky and the air on crisp fall
days, the parks, the places like Frabonis and
the Farmer's Market, the capitol. Another is the
spirit of the places, the sensible, friendly, and
caring way that many things are done.

My first day in Madison was memorable.
I arrived on a Saturday, to register at the last
possible moment. I was working that summer in
Whiting, Indiana for John Gorman, and did not
want to leave. I was having a lot of fun
working there, on such things as nonlinear
estimation. You came breezing through the
department (which, as you recall, was in a house
on Johnson Street) about lunch time and asked
if I had any plans for lunch. I said no, and
you invited me to join you and Gwilym. I then
sat in the back of your VW van as you gave him
a tour of Madison, which included the zoo. You
stopped at El Rancho for some things for dinner,
and, before I knew it, I was having dinner with all
of you. There was champagne, which came out of
the newly opened bottle in an overly vigorous
way, which got everything, tablecloth and all,
wet. Napkins were pushed under the tablecloth,
which gave the otherwise formal setting, a

somewhat casual and definitely lumpy appearance.
As I recall, the champagne was a last-minute
idea as an addition to the menu, and the bottle
was not sufficiently chilled. In any event, a
fine time was had by all, and the evening
went on to songs by you and Gwilym. I think
you both had guitars, and at one point you
were singing statistical songs, impromptu
efforts in rhyme, with you and Gwilym taking
turns. It was a magical day. About 2 am
I left. As I was walking away in the
night I thought to myself, "What a splendid
day. It was wonderful. When I tell anyone
about it, they won't believe it. Neither
will I. I should have a memento. The
champagne bottle is a possibility. That would
be nice to have." I turned around, returned,
knocked on the door. You looked more than
a little surprised when you answered the door,
because being called on at 2 am is quite
unusual. I explained that I'd like to keep
the champagne bottle, and you said fine. That
was the end of my first day in Madison.
   The best thing about Madison is the
friends that I have, which includes Judy,
Jack, and Justin. And you, too, George. I
love you, and I wish you a happy 65th!
                                    Bill.

[CITATION: William G. Hunter, letter to author, n.d.; presented to me on Oct. 12, 1984.]

Here are a few excerpts from letters given to me upon my retirement in 1991:

My decision to undertake graduate studies at Madison in 1960 was pure good fortune. I knew very little about statistics, but I soon learned, thanks to your unique talent for demonstrating how theory and application can support and nurture each other. Your course lectures were delightful, spiced with historical perspectives, personal experiences, and memorable lines from the British music hall repertoire. For me, however, the most valuable lessons were learned outside the classroom, through your willingness to share insights and those memorable Monday evening beer sessions. Perhaps your greatest talent is compelling people to think for themselves.

You have also been an exceptional role model, as teacher, researcher, mentor, consultant and sensitive human being. It was your influence that prompted me to consider a career as a university professor, a career I have found to be enormously satisfying.

DAVID BACON

Your influence in statistics in industry is profound. Your footprints can be found in most of the companies with significant statistical applications through your books, your research, your consulting, your students, and the organizations you created. You have moved the world forward in the application of statistics.

Those of us who studied with you benefit in many different ways. But most of all, we had the opportunity to work closely with one of the greatest minds in the world.

No Man, upon retirement, ever took with him as high a degree of respect and good wishes from so many devoted friends, associates, and students.

GINA G. CHEN

I never thought when I was hired temporarily for 6 months in 1962 it would stretch out to 29 years and still going. We've come through 5 buildings from an old house to a high-rise building, from electric typewriters to computers. I have many fond memories and consider it a privilege to have worked for you and known you.

MARY ANN CLARK

The work that we did together was great fun and, as I hope you know, I've always found your contribution to so many topics (written & spoken) so stimulating, a very special combination of originality and good "sense."

Sir David Cox

As a master craftsman, you have taught so many apprentices over the years. It is wonderful to see the spreading effects of your teachings throughout the world today.

You have done so much for me personally since we first met in 1955 at I.C.I. Blakely, when I came as a summer student. I am grateful for all of it.

Norman R. Draper

When I was a young student I was surprised and delighted that you would include me in your lunchtime walks–to Picnic Point, around Shorewood, around Vilas Zoo–and that you would share so freely with me so many stories about so many things . . .

Thank you for deeply touching both my professional and personal lives, and for your inspiration, guidance, humor, and encouragement shared so generously over so many years. I'm a late starter in many ways, and I have a long long way to go. But now I know the value of Point with Pride, and every day it gets better.

Conrad A. Fung

'It is the duty of the statistician to catalyze the growth of scientific understanding,' you once said in a class. This is, most of all, what I have learned from you. It has been my beacon and my comfort. It has led me to eschew complexity and to demand from myself the effort to penetrate deeply enough into understanding so that the ultimate simplicity ('parsimony:' you would say) that seems to underly most of Nature's doings becomes manifest. It has helped me help others and challenges me to constantly improve.

Bert Gunter

I have learned more from you than I learned from my mother. And my mother was a talker.

Bruce Hoadley

We would like to express, in writing, our strong appreciation for the way we were received and allowed to attend your Monday evening "Beer and Statistics" in your nice house in Madison during our sabbatical year 1987–88. We shall never forget these seminars.

We wish You all the best for Your retirement years, particularly a good health. In that case we feel certain that important contributions to statistics, in particular to the area of Statistical Methods for Quality Improvement, will continue to come from Your hand.

LIV AND ARNLJOT HØYLAND

I also want to thank you for your encouragement of my development as a statistician. You arranged for me to attend and participate in conferences around the country, introduced me to the statistical community beyond Madison, and worked with me on writing technical papers. You have greatly influenced my approach to statistics and problem solving, instilling in me a philosophy and attitude that is both rigorous and practical.

STEPHEN JONES

I continue to marvel at my exceptional good fortune in becoming one of your earliest students. And that's only slightly ahead of being able to call you one of my closest friends. I must have been an awfully good guy in my previous incarnation.

J. STUART HUNTER

Listening to you and observing your interaction with other scientists I have always had the sense that I am in the company of a genius. The four years as your graduate student, and particularly my interaction with you, were instrumental in shaping my life. When I came upon a roadblock in my work I had the sense that you always knew the answer or at least had an intuitive sense about how to proceed.

JOHANNES LEDOLTER

Here are excerpts from some recent correspondence:

We gave a data set (time series) for analysis as a homework problem in one term. ... The model they came up with was not the right one. The right model required a bit more thinking and this became one of the chapters in my thesis. I had the pleasure of reading several papers and projects which came to George for review. He valued my opinions and I appreciated that. One time a question came from a Ford Motor Company engineer about a fractional factorial design and he wanted George to look into it. George asked me to look at it. When I wrote down all the details it became obvious that the standard handbook that they used as reference (published by ASQ or some group) had a typographical error. It was amazing that people were using it without thinking.

BOVAS ABRAHAM

Memories from Gordon Research Conferences on Statistics in Chemistry and Chemical Engineering:

George Box and George Barnard 'engaging in discussion' with Genichi Taguchi following Dr. Taguchi's presentation at a Gordon Research Conference, with the remaining 120+ participants looking on in awe. George Box and Maurice Kendall singing <u>many</u> selections from British music hall repertoire.

George Box mastering the art of using ensembles of smaller and smaller scraps of transparencies for presentations using an overhead projector.

George Box as skit writer (with assistance from Bill Hunter, Svante Wold, Steve Bailey, Dave Bacon and others).

George Box leading Wisconsin alumni in singing "On Wisconsin" at Thursday night parties (once we agreed upon the words).

DAVID BACON

One of the striking things about working with George is that you were his friend first, his colleague second, and finally his student. We were regularly invited to parties just to have fun. There was always music, stories and laughter. And more than a few somewhat bawdy jokes. It was clear that we were part of a team when we were at work. He, like Bill Hunter, had a wonderful way of communicating a vision of how we might help the world by taking a data-based approach to improving the quality of things and the quality of life. Some of his values just pervaded the working atmosphere and we picked them up—a belief that a careful application of the scientific approach can help us understand and improve life, that we shouldn't take ourselves too seriously and should be able to find laughter every day, and that people matter. We mattered to George and he mattered to us. Thanks,

George, for your friendship, guidance and insight. You've made a profound difference in my life and many others.

TIM KRAMER

1) George has always been so generous. Once when we were traveling to a conference he gave me one of his many airline club membership cards so we, the students, could enjoy the club lounge at the airport. So I guess I impersonated George Box for a morning.

2) One time, when my band was giving a concert at the Rathskeller, in the middle of a song I felt a tap on my shoulder. There was George with his big smile cheering me on. He came to the concert with you [Claire] and Judy Pagel just because.

3) George's students (Stephen, Tim and José) would take "field trips" to see George at your home on Branson Road in Oregon, WI. One by one we would go into his office and discuss our research topics, while the other students waited in the living room. Just like being at the doctor's office!

4) I also remember getting phone calls from George early in the morning to go see him. Those were days where George had woken up with an idea that couldn't wait. Sometimes he would give you pieces of paper with some ideas he had jotted down at an airport while he was traveling.

5) At some point in my research he had me look at the ASQC logo, a control chart that looked "under control," and told me that even though the data looked in control, it was not in control in terms of variation. That started the main topic of my dissertation, Cusum charts for variance.

6) And who can forget the parties at your house with music and laughter. I remember one time when George found hats for us to wear. George has always been humorous, a great host, and a great story teller.

JOSÉ RAMÍREZ

I was re-reading *Statistics for Experimenters* the other day and decided to write and thank you for making my life as a Process Engineer much less confusing. I was lucky enough to have attended your 5 day course at the Scandic Crown Hotel in Edinburgh around 20 years ago and what I learned that week has been invaluable over the years. Prior to attending your course, I had been thoroughly confused by a 5 day Taguchi methods course. After attending your course, my confusion was swept away and someone who disliked statistics at school became really interested in it. I just wanted to say a big thank you for helping to make me a much better Engineer than I would have been otherwise.

Mark Taylor
Principal Process Engineer
Fourth Dimension Displays Ltd
Fife, Scotland

Finally, here is a *New Yorker* cartoon from Conrad Fung (by Richard Taylor, *The New Yorker*, March 8, 1952) who wrote the following:

[I] remembered sitting in George's office in 1979 transcribing by hand a manuscript he was working on. His office had two desks facing away from each other at a 90 degree angle; he sat at his main desk and I sat at the other; and as he would finish a page, he would give it to me; and as I would finish a page, I would return it to him. At one point he said, "Have you been taking out my commas?"

"They have a wonderful author–editor relationship."

[Citation: © Richard Taylor/ The New Yorker Collection/ www.cartoonbank.com]

Index

Abraham, Annamma, 137, 204, 205–206

Abraham, Bovas, xxi, 136–138, 164, 204, 248, 261

Adair, John, 178

Ahrend, Jürgen, 224

Air pollution
  Los Angeles, 142

Alice in Wonderland, 9, 145–146, 247

Alonso, Agustín, 239

Altpeter, Roger J., 127

American Society for Quality Control (ASQC), 92, 193

American Statistical Association (ASA), 92
  president of, 106–109, 221

Andersen, Sigurd L., 73–74, 83, 164, 249

Andrews, Horace P., 92

Anscombe, Francis J., 106

Army, British
  Auxiliary Territorial Service (ATS), 33, 35
  balls up, 24
  barrack huts, 27
  demobilization process, 38
  enlistment, 20
  Entertainments National Service Association, 33
  exercises, 24
  explosives, 23, 27, 37
  key man, 20–21
  knots and lashings, 23–24, 27
  old sweat, 23

Army, German, 36–38
  Sergeant Shultz, 37

Army, United States

Army Research Office (ARO), 63
  GIs in England, 25

Arnold, Harvey, 112

Arthur, Mary, 159

Ashkenazy, Vladimir D., 220

Automatic optimization, 125
  self-optimizing reactor at Wisconsin, 125–127

Axley, Ralph E., 148

Bacon, David W., xxi, 96, 164, 249, 258, 261

Bailey, Steven P., 164, 248, 261

Ballentyne, Ford, xxi

Banners (Mr. Marshall), 15

Barnard, George A., 106, 125, 132, 141, 155, 220, 261
  at Princeton University, 56
  brilliant mathematician, 53
  consultant to ICI, 55
  house on Barnes Common, 61
  houseboat trips with Barnard family, 57–58
  initial meeting, 41
  Mill House, 61–62
  office at Imperial College, 53
  Research Committee of RSS, 45
  professor at Imperial College, 41
  professor at University of Essex, 140
  refused visa by FBI, 57

Barnard, Mary, 57–58, 61, 220

Barrios-Zamudio, Ernesto, xxi, 161, 241

Bartorelli's Restaurant, 55

Basu, Asit P., 106

An Accidental Statistician: The Life and Memories of George E.P. Box, First Edition. George E.P. Box.
© 2013 John Wiley & Sons, Inc. Published 2013 by John Wiley & Sons, Inc.

George Box Timeline

1919	October 18: G.E.P. Box born, Gravesend, England
1929	Enters second form at Gravesend County School for Boys on scholarship
1936	Leaves school age 16; becomes Assistant Sewage Treatment Chemist in Gravesend
	Starts attending Gillingham Technical College two afternoons to get coursework to go on for an external degree in chemistry at London University
1938	Takes nine-day intermediate science exam at London University; gains invaluable grounding in scientific method
1939	Begins taking guitar lessons
	June: Publishes first article
	September: War declared with Germany
	October: Ceases studies at London University and enlists in British Army
1939-40	Stationed near Salisbury
1941	Porton Down Experimental Station
	Publishes articles with Professor Harry Cullumbine
1942	Goes to see R.A. Fisher
1945	Marries Jessie Ward
1945	May 8, 1945: V-E Day
	June approx.: Secret mission to German Research Station at Raubkammer for 6 months
	Late December: Demobbed from Army
1946	British Empire Medal
	January: Begins attending University College in London to study statistics under Egon Pearson; completes three-year undergraduate degree with first class honors in 18 months and does graduate work for the remainder of the three years
	Summer: Vacation student at ICI; does work on "Little Davies" and is added as an author
1947	Receives B.Sc., University of London
	Meets George Barnard at meeting of the Royal Statistical Society
	Summer: Vacation student at ICI in dyestuffs Division and receives offer to be salaried there during his third year at university with promise to work there afterward
	Joins ICI's Statistical Methods Panel
1949	Completes three-year education at University College
1949-51	Teaches night class at Salford Technical College using mimeo notes
1950	Begins working on response surface methods (RSM)
1951	Publishes RSM paper with K.B. Wilson
1950s	Attends Professor Maurice Bartlett's lectures at Manchester University
1952	Receives Ph.D.
1953	Goes to North Carolina for a year

	Meets Stu Hunter; Gertrude Cox; Alex Kahlil; etc. — gives a seminar at Princeton and meets John Tukey; goes to his first Gordon Conference and meets Cuthbert Daniel and Frank Wilcoxon
	Learns to drive; buys car; sees the West
	1953–late: Returns to ICI
1954	Writes memo describing EVOP technique to ICI Board of Directors
1955	Meets Norman Draper, a summer student at ICI
1956	Tukey urges George to leave ICI and come to Princeton
	Goes to Princeton to direct the Statistical Techniques Research Group (STRG—Don Behnken, Merv Muller, Henry Scheffé, etc.; begins work with Gwilym Jenkins
1957	EVOP article published
1957-58	Discusses with Stu Hunter and Cuthbert Daniel the need for what would become the journal Technometrics
	—Short course to raise money
1959	Meets Gwilym Jenkins
	Meets Bill Hunter who is a graduating senior at Princeton
	First issue of Technometrics
1959	Comes to Madison to start department; works at Math Research Center
1960	Marries Joan Fisher
	Teaches first statistics classes ("Advanced Theory of Statistics," etc.)
	Bill Hunter begins as a student in the Ph.D. program at the University of Wisconsin
	Stu Hunter in Madison 1960-61
	Gwilym Jenkins in Madison; G.E.P Box and Gwilym receive AFOSR money for a decade
	Meets with Olaf Hougen to discuss automatic optimizer; applies for and receive NSF grant
	Meets George Tiao, Sam Wu
	October: Helen Box is born
1961	Publishes with George Tiao "A Further Look at Robustness Via Bayes' Theorem" Biometrika
	Monday Night Beer Session begins and goes until George's retirement in 1990
1962	May: Harry Box is born
	July 29: R.A. Fisher dies at the age of 73.
	First paper with Gwilym Jenkins, "Some Statistical Aspects of Optimization and Control"
1963	Gwilym suggests that he and George write a book on time series; summers in Lancaster begin
	Bill Hunter finishes Ph.D.; becomes asst prof soon after; assoc in '66; full in '69
	George goes to Indonesia under auspices of Ford Foundation

1964	Publishes "An Analysis of Transformations" with David Cox
1965-66	Spends year at Harvard Business School; works on Bayes book with George Tiao
1967-68	Stu Hunter is first statistician in residence
1968	Statistics Dept now has 17 faculty members
1969	Publishes *Evolutionary Operation: A Statistical Method for Process Improvement,* New York: Wiley, 1969; 2nd ed. 1998 (with Norm Draper)
1970	Publication of first edition of Time Series Analysis: Forecasting and Control with Gwilym, Holden-Day
1970-71	Spends year at Essex; works with George Tiao on Bayes book
1970	Short courses in Spain with George Tiao and Stu Hunter
	Meets Daniel Peña, Albert Prat, Xavier Tort
1973	Publication of Bayesian Inference in Statistical Analysis, John Wiley and Sons, 1973 with G. Tiao
	Begins work on LA auto pollution with George Tiao and Dr. Hamming
1974	Brian Joiner becomes Statistician in Residence; holds position until 1983
1975	Honorary D. Sc., University of Rochester
1976	Second edition of *Time Series Analysis: Forecasting and Control* published by Holden-Day
1978	Publishes *Statistics for Experimenters,* first edition; with Bill and Stu
	President of the American Statistical Society
1980	Deming's broadcast, *"If Japan Can, Why Can't We?"*
1982	George Tiao leaves Wisconsin for Chicago
	George Box goes to Bulgaria (spring)
	July: Gwilym dies
1984	Bill Hunter meets with Mayor Sensenbrenner
	65th birthday party; letters in bound book
	George meets the Queen at the RSS 150th anniversary
1985	January: Don Behnken dies at the age of 60.
	May: Elected fellow of the Royal Society
	Begins the Center for Quality and Productivity Improvement (CQPI) with Bill Hunter
	September: Marries Claire Quist
1986	February: First CQPI Reports issued (nine CQPI Reports issued in Feb. alone)
	June: Trip to Japan to see Taguchi and visit factories (among them Toyota)
	December 29: Bill Hunter dies at the age of 49.
1987	Spring: Taguchi Short Course offered in Madison (GEP, Conrad, Søren)
	Fall: *"Designing Industrial Experiments,"* short course held in Madison for first time (George, Søren, Conrad)
	Publishes *Empirical Model-Building and Response Surfaces* with Norman Draper, New York: Wiley, 1987.
	MAQIN begins

1988	Quality Engineering begins publication; "George's Column" begins
	September: Goes to Sweden to give "Designing Industrial Experiments" short course with Conrad and Søren
1989	Honorary D. Sc., Carnegie Mellon University
	Late May–early June: Trondheim, Norway, gives short course with Søren and Conrad
1990	Videotapes based on short course with Søren and Conrad
1990-91	Spends year at Institute for Advanced Study in the Behavioral Sciences
1991	George retires
	September '91–October '92: Alberto and Marian spend year in Madison-
	ISI meets in Cairo; travels to Israel on same trip
1992	*Bayesian Inference in Statistical Analysis,* second edition, published by Wiley
1993	George and Claire visit Santander, Spain for the first time
1994	Third edition of *Time Series Analysis: Forecasting and Control* by Box, Jenkins, and Greg Reinsel published
1995	Honorary Doctorate, University Carlos III, Madrid
	95–96: Claire and George spend year in Santander
	Early April: Trondheim again; George's talk "The Scientific Context of Quality Improvement"
1997	George and Alberto Luceño publish *Statistical Control by Monitoring and Feedback Adjustment*
1998	98–99: Claire and George live in Santander again
1999	ISI meets in Finland
	October: George Tiao organizes 80th birthday party for George in Chicago
2000	Honorary Doctorate, Conservatoire National des Arts et Métiers (CNAM), Paris
	Honorary Doctorate, University of Waterloo, Canada
2002	August 9: George Barnard dies at the age of 87.
2004	May 5: Greg Reinsel dies at the age of 56.
2006	January 1: Albert Prat dies
2007	Wiley reissues *Response Surfaces, Mixtures, and Ridge Analyses* by George E.P. Box and Norman R. Draper
2009	December 14: Søren dies at the age of 58
	October: Claire organizes George's 90th birthday party in Madison
	Wiley publishes second edition of *Statistical Control by Monitoring and Feedback Adjustment* with Carmen Paniagua-Quiñones as a third author
2010	The University of Wisconsin Department of Statistics celebrates its fiftieth year
	Publishes with Surendar Narasimhan "Rethinking Statistics for Quality Control," which receives a Brumbaugh Award. (George has received this award five times.)
	October: Decides to write light-hearted memoir
2012	Publishes with Bill Woodall "Innovation, Quality Engineering, and Statistics" in Quality Engineering